建筑立场系列丛书 No.6

文博建筑

MUSEUM ART CENTER

韩国C3出版公社 | 编
陈思 徐丽 张琳娜 赵薇 尹航 王单单 董鑫 孙佳 王艺璇 范秀梅 | 译

大连理工大学出版社

04 过程

12 建筑律动 规模上的飞跃

34 博物馆

128 艺术中心

03

温尼伯冰上庇护所

Patkau Architects

温尼伯城位于加拿大大草原上，有将近60万居民。这里冬天持续六个月，是除西伯利亚之外最冷的城市。红河和阿西尼伯恩河在城市中心区交汇，到了冬季人们将雪清理一下便能清理出长达数公里的滑雪道。但是当温度长时间持续在零下三四十度时，想办法找一个防风庇护所将会大大地提高溜冰河道的利用率。

我们的设计提议包含一系列私密的庇护所，每个庇护所一次只能容纳几个人。它们聚在一起形成一个小"村子"（或是"牧群"或是"鱼群"、"鸟群"、"小舰艇队"），构成一个不可分割的集体。庇护所像水牛一样背对着风的方向，仿佛具有生命和目标，聚集在一起保护彼此，抵御寒冷的天气。

每个庇护所都由薄而易弯曲的胶合板构成，胶合板的弯曲变形特性同时满足了结构和空间特性的需要。它的外皮由两层厚度为0.06m的易弯曲胶合板组成，切割成型，固定在由三角形底座、楔形脊柱结构和屋脊构件（屋脊是一条线，抵消了积雪的重力）组成的木质支架上。我们在车间里按照原尺寸做出来了庇护所的原型，然后测算出弯曲面所能承受的压力。这一系列的切口和洞口减轻了压力点的压力。

庇护所的形状是通过压力/变形，然后释放压力的过程得出的。把这些庇护所聚集在一起，一开始只确定其中两个的位置，把它们并列放在一起，但限制入口的大小和可及性。这种表面上随意的搭配实际上是通过120°的精确旋转来实现的。然后把三对这样的组合（一组镜面反射）里的每个小屋都参照临近的一个再旋转90°，按照这种位置摆放成群，在这样较大的组合之中确定"内部"空间。同时，这些庇护所根据特定的方向、一天中时间和环境情况的变化，与日光和风形成了动态的关联。

这些都是精致的、"有生命的"结构。它们在风中轻轻地移动，以不同的频率来回运动并发出咯吱咯吱的声音；它们不稳定地漂浮在结冰的河面上，抖落粘附在结构表面上的积雪。它们易碎纤薄的特性使那些受庇护的人深刻体会到加拿大草原上的冬季不可阻挡、凶猛而又美丽。

Winnipeg Skating Shelters

Winnipeg is a city of 600,000 residents located on the Canadian prairie. It is the coldest city of its size outside of Siberia. Winter can last six months. The Red and the Assiniboine Rivers meet in the center of the city, and in winter, when plowed of snow, skating trails of many miles long are created. But with temperatures that drop to minus 30 and 40 for long periods of time, creating opportunities to find shelter from the wind greatly enhances the ability to use the river skating trails.

Our proposal consists of a cluster of intimate shelters, each accommodating only a few people at a time. They are

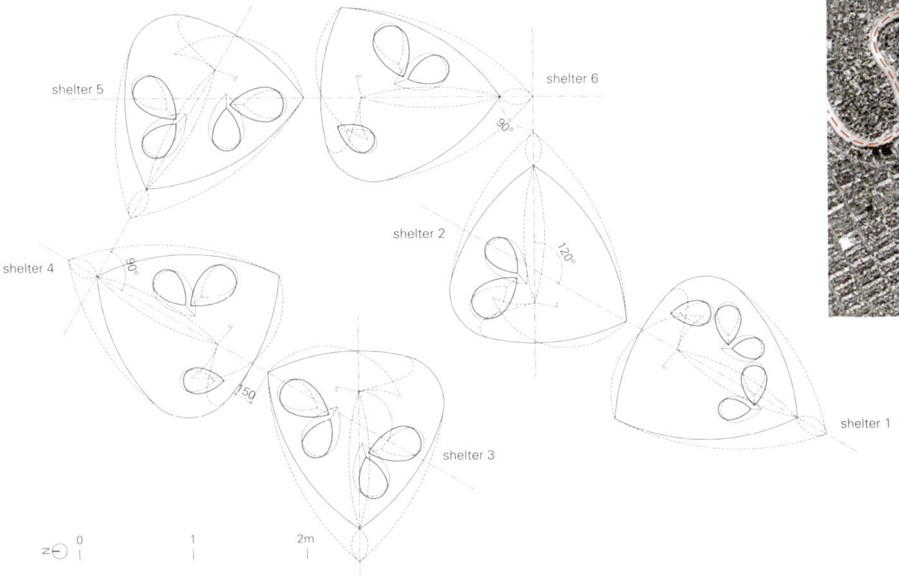
shelter 5
shelter 6
90°
shelter 2
120°
shelter 4
90°
150
shelter 3
shelter 1
N
0
1
2m

6

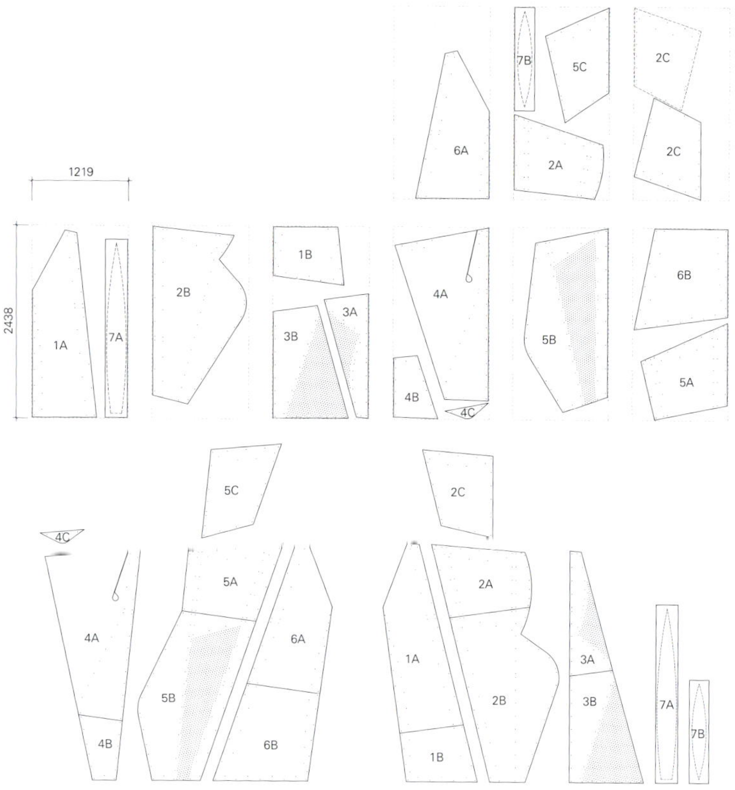

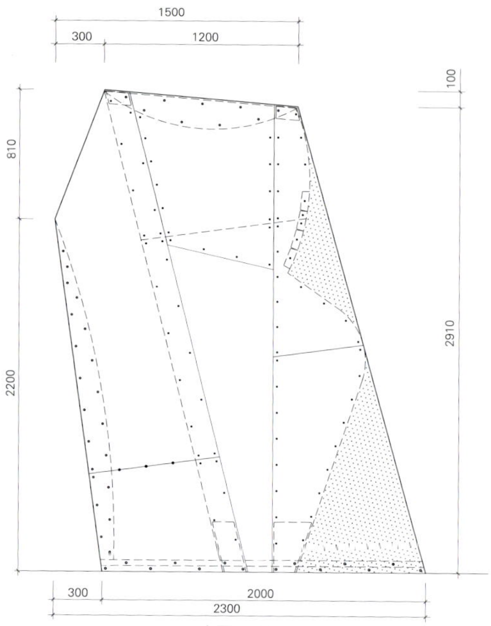

立面A elevation A

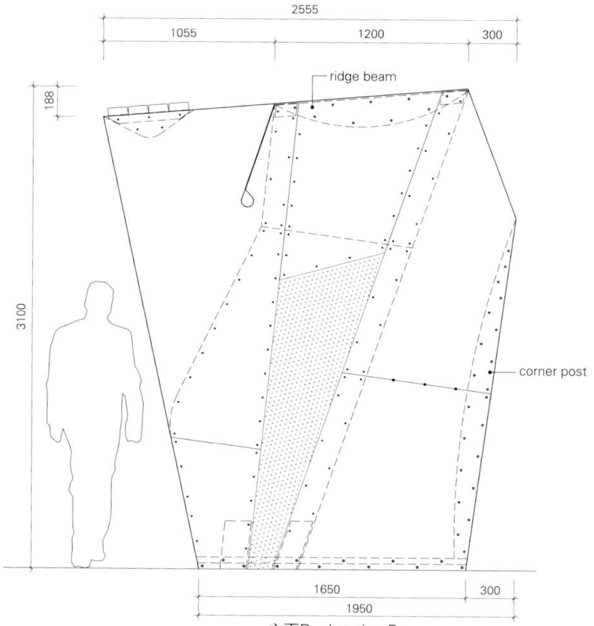

立面B elevation B

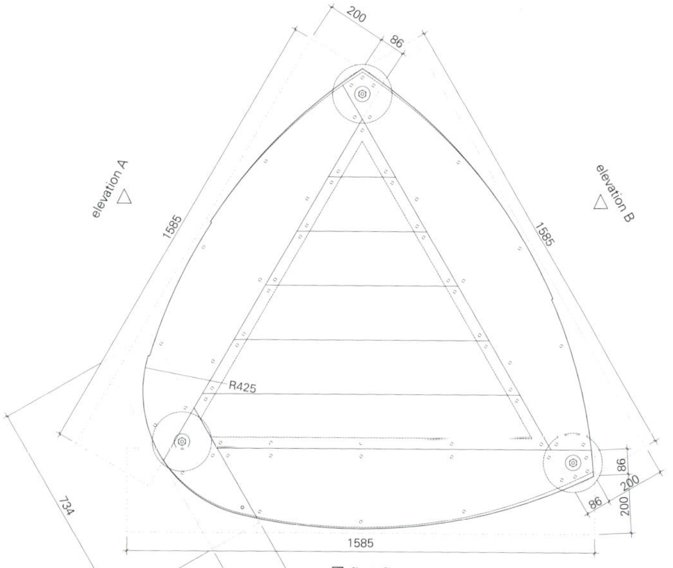

一层 first floor

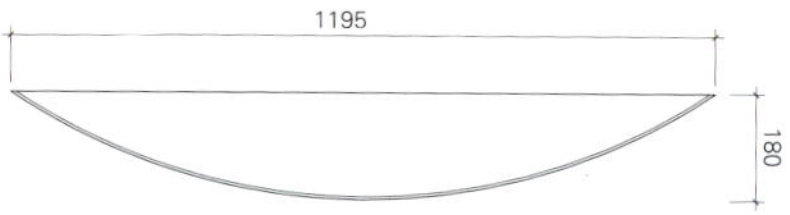

立面1 elevation 1

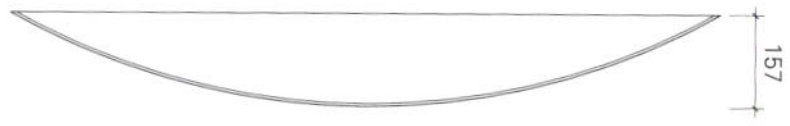

立面2 elevation 2

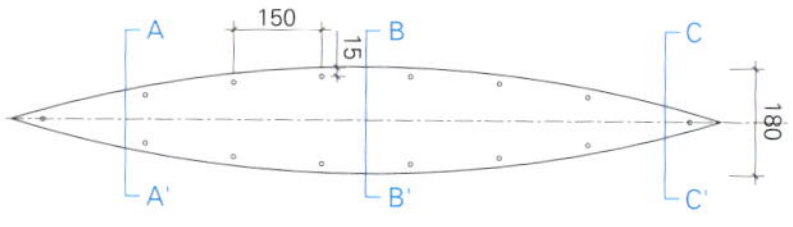

立面3 elevation 3

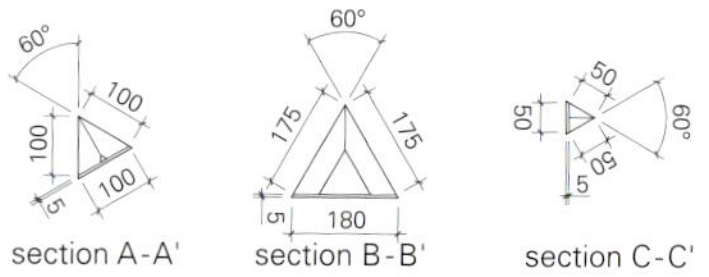

section A-A' section B-B' section C-C'

屋脊梁 ridge beam

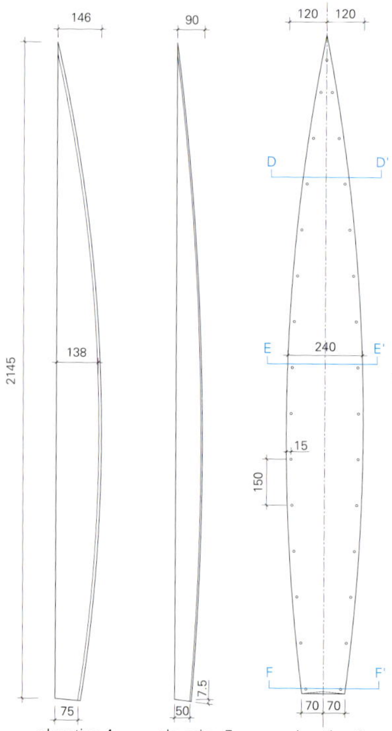

elevation 4 elevation 5 elevation 6

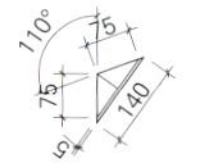

section D-D'

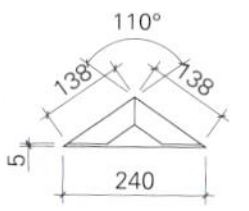

section E-E'

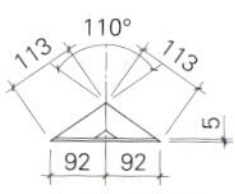

section F-F'

角柱 corner post

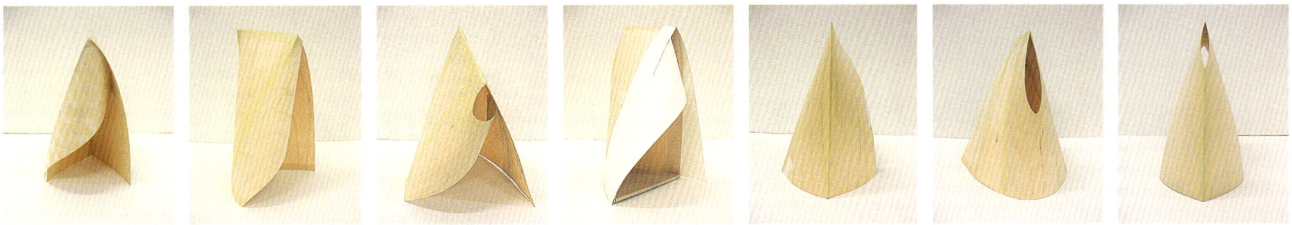

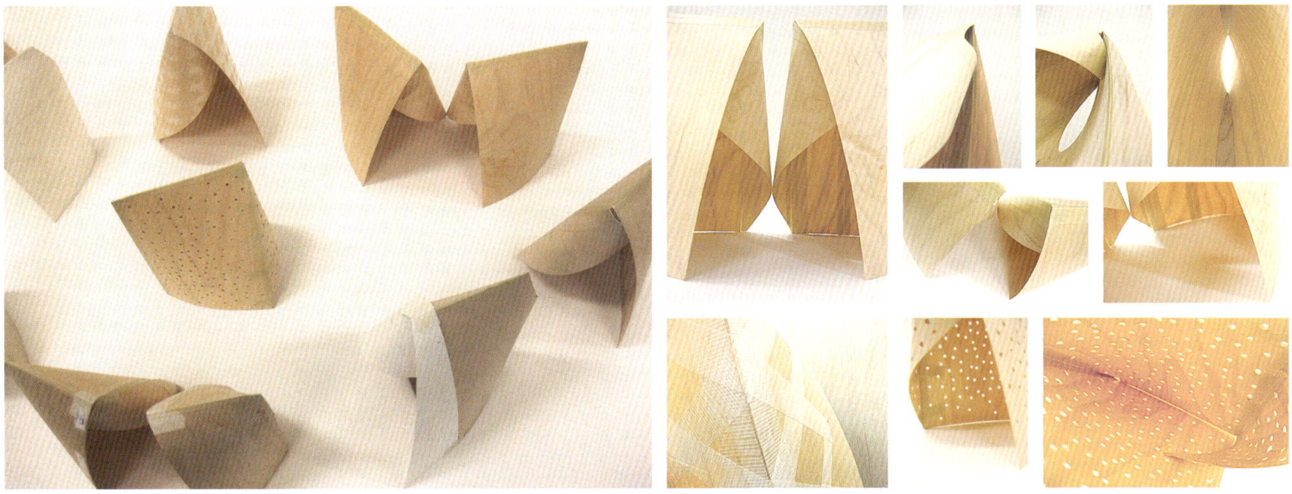

grouped in a small "village" (or "herd", or "school", or "flock", or "flotilla") to form a collective of "something" irreducible to a single interpretation. They stand with their backs to the wind like buffalo, seeming to have life and purpose as they huddle together shielding each other from the elements.

Each shelter is formed of thin, flexible plywood which is given both structure and spatial character through bending/deformation. Skins, made of 2 layers of 3/16 inch thick flexible plywood, are cut in patterns and attached to a timber armature which consists of a triangular base, and wedge shaped spine and ridge members (the ridge is a line to negate the gravity loads of snow). Experiments in our workshop with a full-scale prototype mapped the stresses of bending. Stress points were relieved by a series of cuts and openings.

The form of the shelter is a resultant of this process of stressing/deforming and then releasing stress. Grouping the shelters into a cluster begins with the relationship of two, and their juxtaposition to qualify the size and accessibility of their entrance openings. This apparently casual pairing is actually achieved by a precise 120 degree rotation. Three pairs (one with mirror reflection) are then placed in relation to one another through a secondary rotation of 90 degrees to form the cluster and define an intermediate "interior" space within the larger grouping. Together, the shelters create dynamic solar and wind relationships that shift according to specific orientation, time of day and environmental circumstance.

These are delicate and "alive" structures. They move gently in the wind, creaking and swaying to and fro at various frequencies, floating precariously on the surface of the frozen river, shaking off any snow that might adhere to their surfaces. Their fragile and tenuous nature makes those sheltered by them supremely aware of the inevitability, ferocity and beauty of winter on the Canadian prairies.

项目名称：Winnipeg Skating Shelters
建筑师：Patkau Architects
项目团队：Tyler Brown, Matthew Bunza, James Eidse, John Patkau, Patricia Patkau, Thomas Schroeder, Luke Stern, Peter Suter
合作者：Anna Lisa Meyboom (structural advisor)
地点：Winnipeg, Manitoba, Canada
建筑面积：10m^2 (1.65m^2 x 6 shelters)
竣工时间：2011
摄影师：courtesy of the architect(©James Dow)

自行车架

Manifesto Architecture P.C.

自行车架是人口密集的城市区域力图扩大自行车应用的必要设施。现有的高密度自行车存放系统常常占据了大量的空间，而如果使用垂直升降塔架，主要依靠电力和电脑来操作，结果会导致高昂的经营成本，同时也造成不必要的能量消耗。

自行车架能够存放20至36辆自行车，它提供了一种维护费用低且环保的高密度自行车存放方法，创造出一个像齿轮一样固定自行车的控制站，使用者踩下踏板，就会使存放区旋转起来。此外，自行车架以架台的形式呈现，可以安装在建筑物和保留墙的外墙壁上，更大限度地减少占地面积。这就使自行车架能够把城市中未能充分利用的空间重新利用起来，并且可以把对行人交通造成的阻碍降到最低。这种高于行人的自行车架可以通过可伸缩的钩子，即类似那些安装在汽车后面用于运输自行车的钩子，使自行车架降下来。每一个钩子都设置了用于防盗的锁定装置。而且，即便没有可用的墙壁或空间，直立型的自行车架也可以安装在地面上，可以同时连接多个这样的直立型装置，使它的存放容量无限扩大。

专利审核中的自行车架设计目前已经推广给很多制造商，这样一来，自行车架就能被更多的私人和国家政府机构认识和推广，像韩国、日本、澳大利亚、法国还有德国在内的政府机构，都已经对它表示关注，将它作为在人口密集的城市中心扩大自行车应用的可持续性设计解决方案。

Bike Hanger

The Bike Hanger is an essential facility for dense urban areas seeking to increase bicycle-friendliness. Existing systems of high-density bicycle storage facilities often take up large amounts of space, or if in the form of a vertical tower, they rely heavily on electricity and computerization for operation, resulting in high operational costs and unnecessary energy consumption.

The Bike Hanger, which has the capacity to store between 20-36 bicycles, offers a low-maintenance and environmentally friendly solution to high-density bicycle storage by featuring a control station that functions like a geared stationary bicycle where the pedaling of the user operates and rotates the storage area. Furthermore, rather than taking up valuable public space in plazas and pedestrian passageways, the Bike Hanger takes on the form of a tower

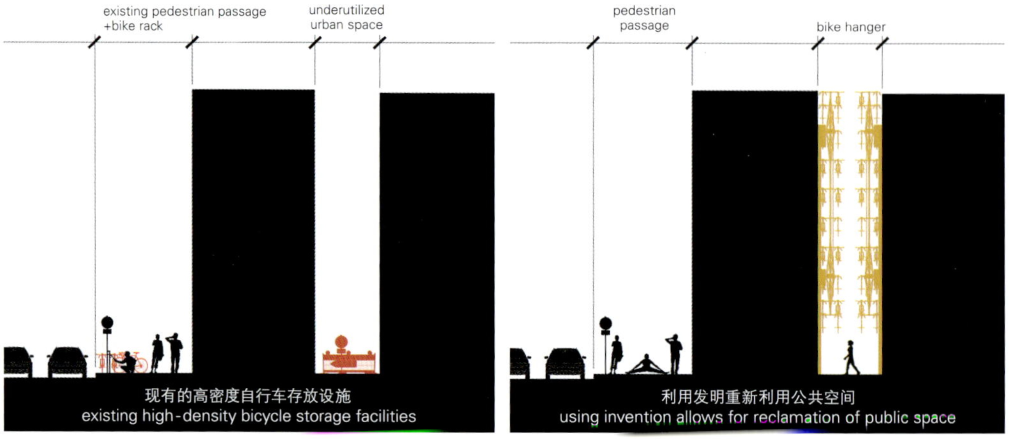

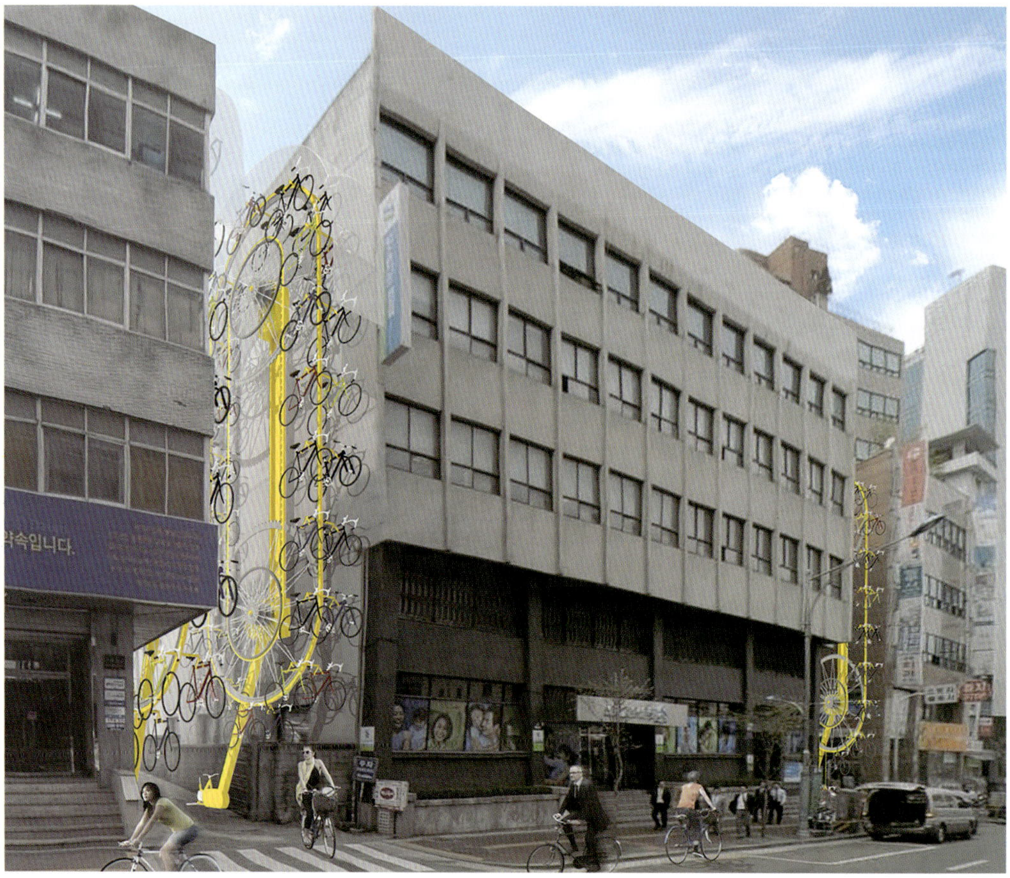

项目名称：Bike Hanger
建筑师：Manifesto Architecture P.C.
项目团队：JeeYong An, SangHwa Lee, Sol Lee, SangHee Jun, JinPyo Eun
建筑面积：7.43m²
设计时间：2010
获奖情况：1st prize winning project

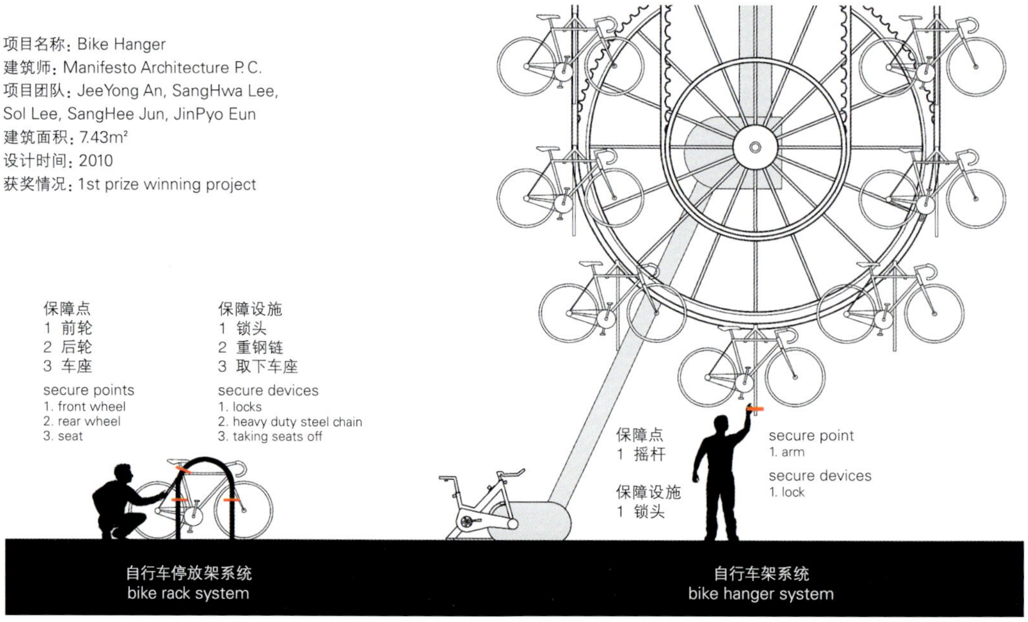

and can be installed on vertical surfaces such as buildings or retaining walls reducing the footprint even more. This creates opportunities for the Bike Hanger to reclaim underutilized spaces that exist around the city and allows for minimal interference with the pedestrian traffic. The bicycles hang above the height of the pedestrians and can be lowered through retractable hooks similar to those used to transport bicycles on the back of automobiles. Each of these hooks features a locking mechanism to prevent theft. Where no wall is present and space permits, a standing version of the Bike Hanger can also be installed on the ground. Multiple units of the standing version can be connected in a series to infinitely increase its storage capacity.

The patent-pending design of the Bike Hanger is currently being introduced to various equipment manufacturers so that it may be realized and delivered to various private institutions and city governments in countries including Korea, Japan, Australia, France and Germany who have shown interest in it as a sustainable and design-conscious solution to increase bicycle-friendliness in dense urban centers.

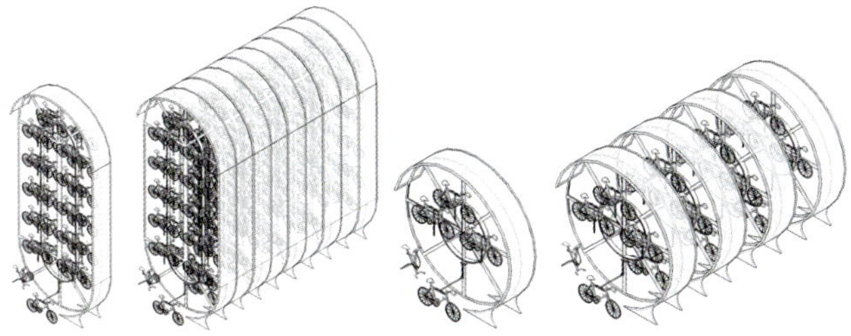

unit 24
H: 10000mm
D: 1100mm
W: 5200mm

unit 120 (24×5)
H: 10000mm
D: 9900mm
W: 5200mm

unit 8
H: 5200mm
D: 1100mm
W: 5200mm

unit 328 (8×4)
H: 5200mm
D: 7700mm
W: 5200mm

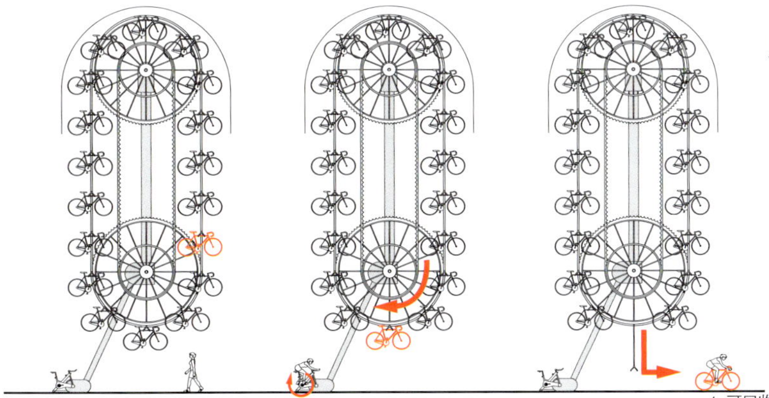

操作图 operation diagram

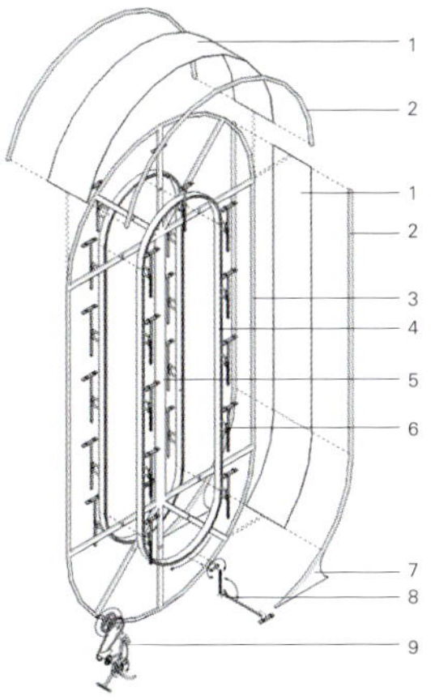

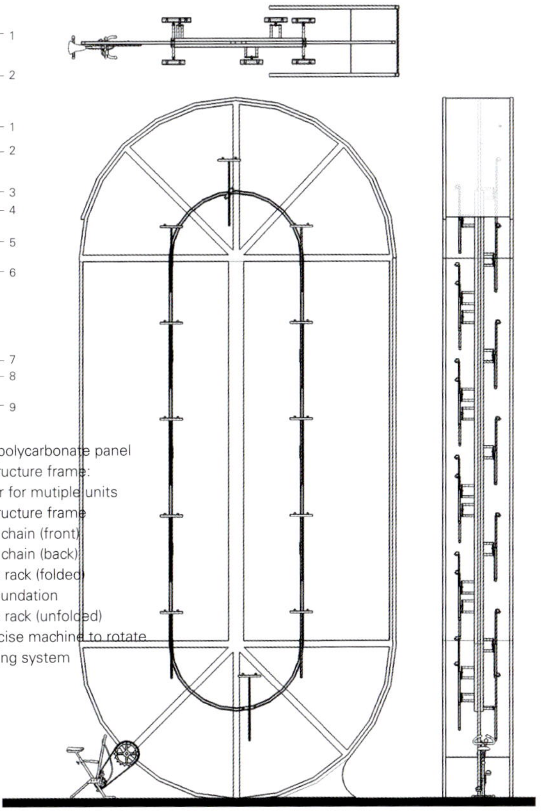

1 可回收利用的聚碳酸酯面板
2 涂漆钢结构框架：
复合构件的连接器
3 涂漆钢结构框架
4 自行车停车链（前方）
5 自行车停车链（后方）
6 钩自行车支架（折叠）
7 车架底座
8 钩自行车支架（展开）
9 自行车运动机器 带动旋转自行车停放系统

1. recycled polycarbonate panel
2. ptd. st. structure frame: connector for mutiple units
3. ptd. st. structure frame
4. bike park chain (front)
5. bike park chain (back)
6. hitch bike rack (folded)
7. hanger foundation
8. hitch bike rack (unfolded)
9. bike exercise machine to rotate bike parking system

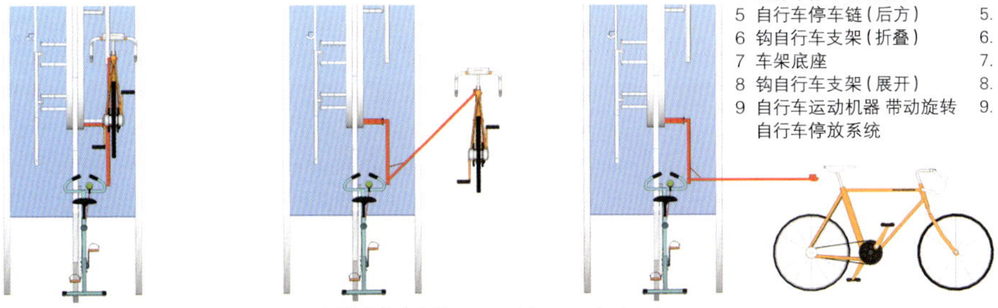

可伸缩的安全钩 retractable security hook

规模上的飞跃——大都市遮阳篷

J. Mayer H. Architects

大都市遮阳篷是由Jürgen Mayer H.带领的柏林的建筑师事务所设计的。它是2003年塞维尔市政当局举办的国际创意竞赛中的获奖方案，该竞赛的目的是为了重振当时的一个“空洞、松散、难以辨认并且功能欠佳的空间”。

这个占地5000m²、由六根支柱支撑的木结构（原先构想为金属结构）充当了一个巨大的遮阳篷，它自身的技术性可以忽视人们的感受，但是当它被置于一个人口稠密的历史城市（像塞维尔市）的中心时，其他与历史相关的问题就会立即显现出来，其中最明显的便是规模问题。

如果一个都市的历史渊源可以追溯到塔西特文明时期，即公元前5世纪的文明，且是由罗马帝国、西哥特王国、摩尔统治和卡斯提尔王国的多种文化融合而成，人们应该对它作何考虑？由于场地空间是人口稠密、错综复杂且建筑高度受限的都市的一部分，人们又当如何将新建的项目嵌入这一大块空间？

人口密度和规模问题

塞维尔市在不忽视流入的古西班牙[1]文化的前提下，更多地强调了其都市生活中长期存在的伊斯兰教文化遗产[2]。伊斯兰教影响该市

Scale Leap

Metropol Parasol

Metropol Parasol – signed by the Berlin-based architecture office led by Jürgen Mayer H. – was the winning proposal of the international ideas competition held by the municipality of Seville in 2003, a contest which sought the revitalization of what was, at that time, an “empty, unstructured, indecipherable and dysfunctional space.”

The technical aspects of this 5,000m² six-legged timber (originally to be metal structure), which acts as an enormous parasol, may in themselves not leave people indifferent, but when an element such as this is placed right in the heart of a dense, historical city center such as that of Seville, other aspects related to History immediately arise, the most evident being the question of scale.

How should one deal with an urban context with origins dating from the Tartessian civilization – back to the V century BC – one which is the result of such different cultures as those of the Roman empire, the Visigoth kingdom, the Moorish domination or the Castilian kingdom? How should one insert the required program within a large empty space that is part of a dense, intricate and height-controlled urban fabric?

The density, scale issue

Without downplaying the inflow of old Hispalis [1], Seville holds more deeply in its urban fabric the lingering legacy of the Islamic presence [2] which persisted in the city for more than five centuries.

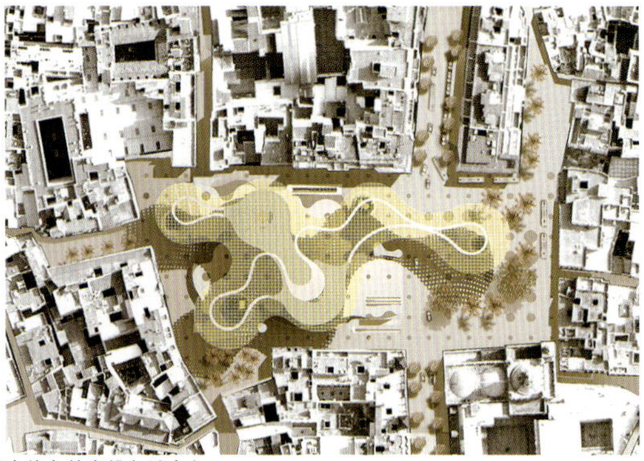

大都市遮阳篷项目嵌入了Encarnación广场，位于人口稠密、历史悠久的塞维尔城中心
Metropol Parasol is inserted in the Plaza de la Encarnación right in the heart of a dense, historical city center of Seville.

至今已逾五百午。

希腊城邦和罗马城市这些古典城市在进行城市设计时都是从内而外、从公共空间（像市场、剧院、戏班、体育场和寺庙）到私人空间的，摩尔时期建的城市却恰恰相反，是从私人空间过渡到公共空间的。它保留了宗教特性，宗教性由个体房屋表现出来——因为穆斯林们把私人住宅看作某种神圣之地，其中的私人生活不能受打扰[3]——这一点贯穿了所有方面。这些城市都有私密、与世隔绝和神圣的方面，意味着人们可以居住其中而城市本身并不浮夸[4]。通常来说，我们称作街道的地方只不过是私人区域之间留存的空间罢了；公共空间在进行城市设计时却失去了自身的结构作用。

在塞维尔，伊斯兰式城市的复杂而私密、隐蔽、不加区分且拥挤的布局仍然在延续，甚至在基督教统治时期，许多地区的种族少数派，尤其是摩尔人和犹太人这些大都被孤立在飞地里的人们，仍然对这种布局表示支持。文艺复兴的清风并没能在塞维尔找到合适的地方来实践它的对称与透视的思想，但是后来法国大革命的反圣职主义意识形态却使先前不可想象的活动得以实现。在约瑟夫•拿破仑•波拿巴[5]王国统治的那几年，法国工程师萌生了许多都市革新想法，他们拆毁了许多修道院和教堂，想为都市生活创建出一些开放空间。Encarnación广场就是那些年他们修建的广场之一，它的名字来自于1810年被拆毁的一个古老的修道院。而大都市遮阳篷项目就是嵌入了这个广场中。

五六层高的涂白的建筑连续统一体以古老的穹顶和塔楼作点缀，又间或有百年古树荫庇下的矩形广场作调节，描画出的塞维尔市中心的天际线现今被许多波状的结构体打断了，这些结构体倾斜着伸出建筑顶部露台和阿拉伯式的瓷砖屋顶。当一个人走过附近人口稠密且狭窄的街区时，他们断断续续的、简短的视线现在被上方一个巨大、波状、交叉的物体打断，或说被几根支柱支撑起来的遮阳篷打断。它的魅力给人留下深刻印象，并且在整个都市组成部分中起着主导作用。

大都市遮阳篷与大教堂——这个巨大的历史城市中心的最大建筑——的规模类似，它的表面在原有建筑上方不远处起伏铺延，与该市古老的塔楼（吉拉尔达塔和黄金塔）所代表的数世纪历史相抗衡。参数设计超越了传统，形成了自己的形式和机制，向现存的设计规则提出挑战。曲线、交织和并行的面主导了18 000m^{2}的广场空间。

Contrary to what happens in classic cities – the Greek polis and the Roman civitas – in which the city is planned from the outside to the inside, from the public spaces (the agora, the theater, the circus, stadiums, temples) to the private, the Moorish city is the result of the prevalence of the private space over the public. It maintains its religious character, which arises from the individual house – considered by Muslims a sort of sacred space in which the private life is not to be disturbed [3] – and permeates everything. The private, hermetic and holy aspects of these cities mean they are to be lived in without making a show of themselves [4]. For the most part, what we might read as streets are merely the spaces remaining between private domains: the collective space has all but lost its structural role in the design of the city.

In Seville, the intricate, intimate, recondite, undifferentiated and dense urban layout of the Islamic city continued, even during Christian rule, sustained in many areas by the presence of ethnic minorities, especially Moors and Jews, who were generally isolated in enclaves. The fresh air of the Renaissance did not find in Seville an apt ground in which to draw its ideas of symmetry and perspective, but the anti-clericalism of French Revolution ideology would later allow for the execution of previously unthinkable actions. In the years of the kingdom of Joseph-Napoléon Bonaparte [5], the innovative urban ideas of French engineers led to the dynamiting and destruction of many convents and churches, with the idea of creating open spaces within the urban fabric. One of the squares created during those years is the Plaza de la Encarnación, which takes its name from the old convent demolished in 1810. It is on this site that the Metropol Parasol project has been inserted.

The continuum of whitish plastered five-to-six-storey buildings – mottled with historical domes and towers, occasionally relieved by rectangular squares shaded by centennial trees – the skyline that characterizes the large Seville city center – is now disrupted by a wavy mega-structure that leans out over the building-top terraces and Arabic tile roofs. The broken, short perspectives that one gets while walking through the dense, narrow neighborhood streets are now interrupted by a large, wavy, cross-linked presence that top-frames one's view, or by one of the gigantic legs supporting the parasol, which looms impressively, assuming leadership over the entire urban composition.

With a footprint similar to that of the cathedral – the biggest building of the large historical city center – the Metropol Parasol waves its surfaces slightly over the existing buildings, entering in competition with centuries of history represented by the old towers of the city, la Giralda and the Torre del Oro. The parametric design

建筑师Fernando Alda 提供
courtesy of the architect (©Fernando Alda)

2011年3月在建图
under construction in March 2011

建筑师Fernando Alda 提供
courtesy of the architect (©Fernando Alda)

J. Mayer H.简要地介绍了这个项目：

"大都市遮阳篷基于考古学角度深入塞维尔的历史，同时又涵盖了21世纪新都市环境中的城市核心。该项目涵盖了历史，还容纳了食品市场与商业空间的日常活动，并且为重大事件与沉思冥想提供了公共空间。它拥有上升的结构，涉及到图标文化，目的在于使参观者看到塞维尔的屋顶景观的全貌。所有这些不同的程序在白天和夜晚的多个时间都是开放且活跃的。"

格架遮阳篷本身便是主导元素，它在塞维尔夏季的高温里为广场提供荫凉。原先设计是一个覆有穿孔铝构件的轻质金属屋顶结构，后来因为这种尺寸的金属有建造的限制，才改用大型的、以聚氨酯覆膜的木材建造。这一新的结构被认为是当今规模最大的木结构工程之一，由3400块独立的木材和3000个承重能力强的节点构成。"塞维尔作为西班牙一个较大的城市，就像一个不夜城。无论何时在城中漫步，公共广场总有活跃的氛围。与阳伞白天遮阳的作用差不多，大都市遮阳篷为多种形式的公众活动提供了一个具有艺术氛围的场所，在夜晚甚至扮演着更为重要的角色。"建筑师强调说。

J.Mayer对项目背景与塞维尔市中心城市历史建筑规模之间的联系作了阐述：

"项目的出发点是为广场提供荫庇，来营造一种"气候"，即使是在炎热的夏季仍能吸引游客。我们四处寻找一些有趣、特别且能进行抽象化和探索的东西。规模对我来说是个非常重要的概念。我们经常改变场地的各种元素来改变规模，所以在某种程度上，我们再看我们的项目，就像是通过特定的镜头在周围的背景下用新鲜的眼光看到新鲜的东西一样……我们参考了邻近的del Cristo de Burgos广场的大树和大教堂波状起伏的石质屋顶。"

新的独立元素遵循了其自身的形式规则，并象征着在人口稠密的古老城市环境中的一次规模大飞跃。这一飞跃也使我们对某些建筑图标价值进行反思。

图标问题

1992年塞城举办了一次世界博览会，这次博览会是为了庆祝发现美洲大陆500周年，六个月内吸引了超过四千万游人参观游览。自此，塞城在当代建筑领域内巩固了自身地位，比如卡拉特拉瓦设计的阿拉米罗大桥，这一类项目就强化了古老塞城的当代风貌。

此外，20世纪90年代后期以来，我们见证了市政当局保护城市

transcends the tradition and establishes its own conditions and mechanisms, challenging the existing tissue. Curves, twists and parallel surfaces take over the 18, 000 m² of the square space.

J. Mayer H. briefly explains the project thus:

"Based on an archaeological window into the history of Seville, Metropol Parasol covers the very heart of the city for 21st century's new urban place. The project covers history, hosts the everyday life in the food market or in commercial spaces, offers open public space for events and contemplation, and refers to a visionary culture with rising structures to elevate visitors to a panoramic view onto the roof landscape of Seville. All these different programs are open and active at various times of day and night."

The dominant element is the latticed parasol itself, which shades the square during the high summer temperatures of Seville. What was originally to be a lightweight metal roof structure covered with perforated aluminum elements was changed to a large polyurethane-coated timber construction due to restrictions on metal construction in these dimensions. The new structure, considered one of the world's largest timber engineering constructions, has 3,400 individual wooden elements with 3,000 high-load-bearing connection nodes. "Seville is, as most bigger cities in Spain, very close to a 24-hour urban space. Whenever you walk around in the city, there is a lively atmosphere. As much as the parasols provide shadow during the day, nighttime might become even more important when Metropol Parasol creates an atmospheric cover to various forms of public activities still to be invented," the architect emphasizes.

J. Mayer explains the relation of the project with the context and the firm's approach to the historical urban tissue of the city center of Seville as a matter of scale:

"The starting point was to bring shade to the plaza, to create a climate that invites people even in hot summer months. We looked around to find something interesting and specific that could be abstracted and explored. Scale is a very important concept to me. We often shift scales of various elements found on site. So in a way we see our projects as certain lenses through which the surrounding context is looked through to see something new with a fresh eye. [...] Our references are the big trees on the neighboring Plaza del Cristo de Burgos and the undulated stone roof of the cathedral."

The new autonomous element follows its own formal rules and represents a considerable scale leap in the dense and historical urban tissue. This leap points the way to a reflection on the iconic value of certain architecture productions.

Iconic issue

After the Universal Exposition was held in Seville in 1992–orga-

David Baron 提供
courtesy of David Baron

毕尔巴鄂的古根海姆博物馆，1997年，弗兰克•盖里
The Guggenheim Museum Bilbao, 1997, Frank O. Gehry

Graeme Maclean 提供
courtesy of Graeme Maclean

瑞士再保险总部大楼，伦敦，2003年，Foster+Partners
Swiss Re HQ, London, 2003, Foster + Partners

现存历史建筑的意图，并基于建筑的体量、建筑形式或建筑材料等，把这些因素应用到建筑设计中，使塞城有别于其他城市。这一举动就像打响了一场建筑竞赛的发令枪。一些项目，诸如众所周知的毕尔巴鄂的古根海姆博物馆，以及伦敦所谓的“腌黄瓜”塔，北京的国家体育馆都是很明显的例子。塞维尔通过大都市遮阳篷这一项目明确表示决不逃避竞争。

是技术上的夸耀？还是城市的新图标？大都市遮阳篷似乎预见了未来塞维尔市的其他建筑，如阿根廷籍美国建筑师C.Pelli设计的178m高的42层塔楼，又如新近完成的[6]——高度上不是那么张扬——由R.Rogers设计的帕尔马斯阿特拉斯科技综合楼。

J. Mayer H.建筑师事务所关注的重点是建筑、交流和新技术交汇处的作品，他将人体、科技和自然之间的关系作为新空间产生的背景。这也解释了选择他的提案作为竞赛的获奖作品的理由：“建筑是一种催化剂，它不是日常生活的背景，而是促进你对空间重新思考的物体。我经常领略到人们的好奇心与自尊心，并且我相信这些情绪会驱动从戏剧到舞蹈，从时尚到艺术等各种层面上的文化产品的生产。这是探索的契机。”

根据查尔斯•詹克斯的图标建筑理论[7]，建筑应该是具有多重编码，并可以进行多重解释的。建筑的外形要能够激发人们的想象，这样我们就可以用隐喻的手法对它们进行描述。这座城市倾向给自己最出众的建筑[8]起绰号，大都市遮阳篷也不例外，在它今年三月中旬向公众开放前，就有了它的绰号——Las Setas（蘑菇）。

项目的含义

建筑师阐述了设计背后的原因和程序上的原理：“大都市遮阳篷汇聚了我们事务所在近些年涌现出的大部分概念和理念。我们尝试着将参数作为每个项目的骨架或框架。这些都是基于甲方的理念、对文脉的参考和功能性逻辑的概念性作品，而不是受设计启发而产生的艺术作品。”

建筑师所指出的项目关键——受概念而非设计启发的作品——暗示出其重点在对个体的规划和对具体事物的策划上。考虑到建筑复杂的背景，包括人口密度、历史、传统、城市文脉和创新、新奇感和图标价值，J. Mayer阐释道：

西班牙的建筑被认为是一种伟大的文化冒险——如果你仔细审视一下他们丰富而富有创造性的建筑历史，你就不会感到惊奇。因此，

nized to celebrate the 500th anniversary of the discovery of the America and visited by 40 million people in six months–the city strengthened its position within contemporary culture. Projects such as the Alamillo bridge, signed by Calatrava, began to emphasize the contemporary face of historical Seville.

Moreover, since the late 90's we have witnessed a growing intention of municipalities to secure a place for their cities within contemporary architectural history, marked by an increased willingness to erect buildings designed to distinguish the city from others, based on their size, or form, or materialization, etc. The starting gun has sounded for the iconic architecture race. Projects such as the well-known and widely published Guggenheim museum in Bilbao, the so-called Gherkin tower in London, or the Olympic stadium in Beijing are clear examples. Seville, as evidenced by the Metropol Parasol, has obviously decided not to hold itself above the fray.

A technical boast? A new icon for the city? Metropol Parasol seems to be a project foreshadowing others in Seville, such as the coming 42-floor, 178-meter Cajasol tower signed by Argentinian-American architect C. Pelli, or the recently finished[6]–and much more restrained in height–Palmas Altas Technology Complex by R.Rogers.

J. Mayer H. Architects, who focuses on works at the intersection of architecture, communication and new technology, and who conceives the relationship between the human body, technology and nature as the background for a new production of space–interprets the selection of his proposal as the winning project of the ideas competition as follows: "Architecture is a catalyst, which is not a background to an everyday life, but something that provokes you to rethink spatial conditions. I see a lot of curiosity and pride, which I believe drives all levels of cultural production, from theater to dance, to fashion and art. It is a moment of exploration."

According to Charles Jencks' theory of iconic buildings[7], they should be multi-codable and multi-interpretable, and their shapes should stimulate the imagination so that we can project metaphors onto them. In a city rather prone to supplying nicknames for its most distinctive projects[8], the Metropol Parasol has been no exception, having acquired the nickname Las Setas (The Mushrooms) even before it was open to the public in the middle of this March.

Implications of the project

The architect explains the reasoning and procedural method behind the design: "Metropol Parasol brings together most of the concepts and ideas boiling in our office for some years now. We try to establish parameters as a skeleton or framework for each project. These are conceptual conditions rather than design driv-

Mark Bentley 提供
courtesy of Mark Bentley

帕尔马斯阿特拉斯科技综合楼，毕尔巴鄂，2009年，Rogers Stirk Harbor+Partners
Campus Palmas Altas, Bilbao, 2009, Rogers Stirk Harbour+Partners

Francisco Diez 提供
courtesy of Francisco Diez

北京国家体育馆，2008年，Herzog & de Meuron
Beijing National Stadium, 2008, Herzog & de Meuron

我迫不及待地想要看看我们对交流和公共空间的猜想将会如何改变塞维尔其他的重要建筑并与其相互作用。现在，大都市遮阳篷项目已经交付使用，开始了它自己的生存历程。

大都市遮阳篷成为了许多报纸文章报道的对象。一些报道对其未来的景象刊出了溢美之词，而一些报道因这个备受争议的项目最终的预算之高和交付时间的拖延而对其表示抵制。但有一点是值得肯定的，就是大多数人都对它产生了兴趣。

矛盾的是，大都市遮阳篷坐落在Encarnación广场[9]上，它已经被描述为这座城市的新标志，它是否是这座历史城市意识到必须要面向新时期的体现？它是否代表了新塞维尔的化身？这期间的跨越有多大？只有时间能说明一切。

1 罗马城市的名称

2 从711年开始，穆斯林对伊比利亚半岛的统治长达8个世纪。它的占领将基督教的领地边界推到了北部的坎塔布连山。但后来形势逆转，天主教国王最终于1492年征服了格拉纳达，反抗成功，重新收复了伊比利亚半岛。

3 《可兰经》第四十九章有助于人们了解穆斯林对住宅的看法。第四节：那些在（你和你的妻子专用的）私人房间后边大声喊叫的人，大多数是无知的人（因此，他们是没有礼貌的）。第五节：如果他们能够耐心等你出来，则会更好。不过安拉通常是宽恕的、仁慈的。

5 在巴约讷退位、拿破仑军队入侵以后，约瑟夫•拿破仑•波拿巴由他的弟弟，即法国国王，指派为西班牙的国王，于1808—1813年进行统治。

6 在大都市遮阳篷竣工之前，也就是它在竞赛中胜出的两年之后，帕尔马斯阿特拉斯科技综合楼成为2005年举办的竞赛的获奖作品。

7 《图标建筑》，Cf. Jencks, C., Frances Lincoln Ltd, London, 2005年

8 例如，V Centenario大桥因为与旧金山的金门大桥很相似，但是规模要小得多，被昵称为"帕基托"或小"柏高"（旧金山的昵称）。其他广为人知的建筑的绰号是因为与名人或邻近乡村类似而得来的。

9 "化身"的西班牙词语，意思是代表或具有某种特质、理念等的人或事物。

en compositions, based on a client's brief, contextual references and programmatic logistics."

The architect supplies the key to the project–"concept rather than design driven compositions"–indicating an emphasis on the outline over the particular, the scheme over the concrete. Conscious of the delicate ground of such notions as density, history, tradition, and urban grain as well as of innovation, surprise and iconic value, J. Mayer reflects:

Architecture in Spain is considered a great cultural adventure–actually not surprising if you look at their rich inventive architectural history. So I am very curious to see how our speculations about communication and public space might transform and interact with other important projects in Seville, now the project Metropol Parasol is handed over and begins its own life.

The Metropol Parasol has been the subject of many newspaper articles, some offering congratulations on the coming spectacle, some rejecting a controversial project based on its final budget or the delay in its delivery. What is certain is that it does not leave many people apathetic.

Paradoxically situated in the Plaza de la Encarnación[9], will the Metropol Parasol–already being described as the new icon of the city–be the embodiment of a historical city's awareness of the necessity of looking toward new periods? Does it represent an incarnation of the new Seville? How big, actually, is the leap? Only time will tell. Marta González Antón

1. The Roman name of the city.
2. The Iberian Peninsula was under Muslim rule for almost eight centuries, beginning in the year 711. Its occupation pushed the bounds of Christian domination up to the Cantabrian north; the inverse movement of reconquest was completed in 1492, with the final conquest of Granada by the Catholic Kings.
3. Chapter 49 of The Holy Quran offers a hint of the Muslim understanding of the house. Verse 4: "Those who call out to you from behind the private apartments (which you share with your wives), most of them do not reason and understand (and are, therefore, lacking in good manners)". Verse 5: "And if they had patience till you could come out to them, it would have been better for them. And Allah is Oft-Forgiving, Most Merciful".
4. For an overview, see Chueca Goitia, F., "Breve historia del urbanismo", Alianza Editorial, Madrid, 1968.
5. Joseph Napoleon Bonaparte was appointed King of Spain by his younger brother, the French Emperor, following the Abdications of Bayonne and the invasion of the Napoleonic army. He reigned from 1808 to 1813 as José I Bonaparte.
6. Finished before the Metropol Parasol, the Technology Complex of Palmas Altas is the result of a competition launched in 2005, two years after the contest which begat the former.
7. Cf. Jencks, C., The Iconic Building, Frances Lincoln Ltd, London, 2005
8. The V Centenario bridge, for example, has acquired the nickname of "Paquito," or little Paco (a friendly rendering of "Francisco"), due to its similarity – at a much smaller scale – to the well-known Golden Gate bridge in San Francisco. Other projects' nicknames widely used stem from comparisons to celebrities or neighboring villages.
9. The Spanish word for incarnation, embodiment: A person or thing that typifies or represents some quality, idea, etc.

Marta González Antón，建筑师，现住在荷兰鹿特丹，并在那里工作。她曾在西班牙和意大利求学。在去往鹿特丹之前，她与西班牙多家当地事务所合作，并从事出版工作。她现在对荷兰和西班牙两国之间的现代设计进程进行研究。

Marta González Antón, architect, works and lives in Rotterdam, the Netherlands. She studied in Spain and Italy. Before moving to Rotterdam she collaborated with several local offices in Spain and worked on publication activities. She is currently carrying out a project research about the contemporary design process between the Netherlands and Spain.

项目名称：Metropol Parasol
建筑师：J. Mayer H. Architects
地点：Seville, Spain
用途：archeological site, farmers market, elevated plaza, multiple bars and restaurants
基地面积：18,000m²
建筑面积：5,000m²
总建筑面积：12,670m²
建筑规模：four stories above ground
设计时间：2004~2005
施工时间：2005~2011
造价：EUR 6,500,000
摄影师：courtesy of the architect-p.20(except as noted), p.30
©David Franck-p.18~19, p.21, p.25, p.26~27
©Roland Halbe-p.12~13, p.22, p.24, p.28~29, p.32

大都市遮阳篷是塞维尔的一个新标志。它彰显了塞维尔作为西班牙最有魅力的文化胜地之一的形象，推动了Encarnación广场成为塞维尔新的城市中心。它独特的空间与塞维尔中世纪的内城紧密结合起来，适于进行各种各样的活动，例如思考、休闲和商业活动。高度发展的基础设施为广场增添了活力，使其成为当地人和游客钟爱的目的地。该遮阳篷从考古学角度深入塞维尔的历史，同时又涵盖了21世纪新都市环境中的城市核心。该项目涵盖了历史，还容纳食品市场与商业空间的日常活动，并且为重大事件与沉思冥想提供了公共空间。同时凭借其上升的建筑结构为参观者提供了眺望塞维尔梦幻般文化全景的地方。这些项目在白天和晚上的不同时间开放，遮阳篷在白天为人们提供荫凉，随着它举办更多渲染气氛的公共活动，晚上的时光会变得越来越重要。

大都市遮阳篷设计有巨大的结构，在底部和内部设有一个考古场地、一个农贸市场、一座广场、多元酒吧和餐馆，在顶部还设有可观看城市全景的露台。创新性的木质结构喷上了聚氨酯，遮阳篷从文化古迹胜地成长为现代化的里程碑。梁柱成为进入下层博物馆和支撑上面全景平台的突出标志，诠释了历史性和现代化城市的独特关系。

大都市遮阳篷成为赋予Encarnación广场和塞维尔新的现代身份的主要建筑元素，设计者将它构思为一个高大的屋顶结构，在白天，它能够提供奇异的阴影，并且创造了舒适的微气候区和娱乐场所。在傍晚，它成为一个为各种灯光和声响方案提供舞台的人造天空。这两个特征使其能够进行一系列的城市活动（例如水上排球、拳击、街头篮球）和商业用途（汽车和服装展览、公司活动、发布会），这突出了广场作为城市主要的交流和互动场所的重要作用。

高大的露天广场满足了在塞维尔古老的市中心举行盛大的、多功能集会的需要。它由巨大的阶梯通向街道区域，为街道生活增添了活力。阶梯为通往音乐会、剧院演出和服装展览等限制性入口提供再造格局，标志着广场成为足够大的公共聚集地。从视觉上讲，平台上的宾馆与下面的市场和地面上的博物馆相连。这种建筑方案将原始的区域扩大一倍，创造了4500m²高质量的节日表演区。位于表面特定区域的水池进一步增强了微气候区功能，并且提供坐和停留的地方。

经过几个月对结构可能性的探索，我们决定以曼沙工程作为原型，大都市遮阳篷也将由薄的木板建成，表面涂上聚氨酯。较低的部分将用钢和混凝土建成。由于防火的规定，广场的就餐区域设有混凝土通风井和钢质结构。但是，广场本身是木质结构。考虑到很复杂的因素，例如预制、维修、花费、生活距离、火势、地震、交通的负荷量，及气温动态等，我们认为木制品是胜任此项要求的最佳选择。

大都市遮阳篷重新界定了Encarnación广场成为塞维尔城市中心的潜力，这里建筑师提出的建筑方案不仅为塞维尔的市民提供了新的地域形象，而且展现了塞维尔作为西班牙最有吸引力的目的地的重要地位。

Metropol Parasol

Metropol Parasol is the new icon project for Seville – a place of identification to articulate Seville's role as one of Spain's most fascinating cultural destinations. Metropol Parasol pushes the potential of the Plaza de la Encarnación to become the new urban center of Seville. Its role as a unique urban space within the dense fabric of the medieval inner city of Seville allows for a great variety of activities such as contemplation, leisure and commerce. A highly developed infrastructure helps to activate the square, making it an attractive destination for tourists and locals alike. Based on an archeological window into the history of Seville, the parasols cover the very heart of the city for a new urban place for the 21st century. Metropol Parasol covers history, hosts the everyday life in the food market or in commercial spaces, offers open public space for events and contemplation, and refers to a visionary culture with rising structures to elevate visitors to a panoramic view onto the roof landscape of Sevilla. All these different programs are open and active at various times of day and night. As much as the parasols provide shadow during the day, nighttime might become even more important when Metropol Parasol creates an atmospheric cover to various forms of public activities still to be invented.

The Metropol Parasol scheme, with its large structures, offers an archeological site, a farmers' market, an elevated plaza, multiple bars and restaurants underneath and inside the parasols, as well as a panorama terrace on the very top of the parasols. Realized as an innovative timber-construction with a polyurethane coating, the parasols grow out of the archeological excavation site into a contemporary landmark. The columns become prominent points of access to the museum below as well as to the plaza and panorama deck above, defining a unique relationship between the historical and the contemporary city.

The Metropol Parasol is the main architectural element that gives a new contemporary identity to the Plaza de la Encarnatión and Seville as a whole. It is conceived as a roof structure that provides

建筑师（Ignacio Ysasi）提供
courtesy of the architect (©Ignacio Ysasi)

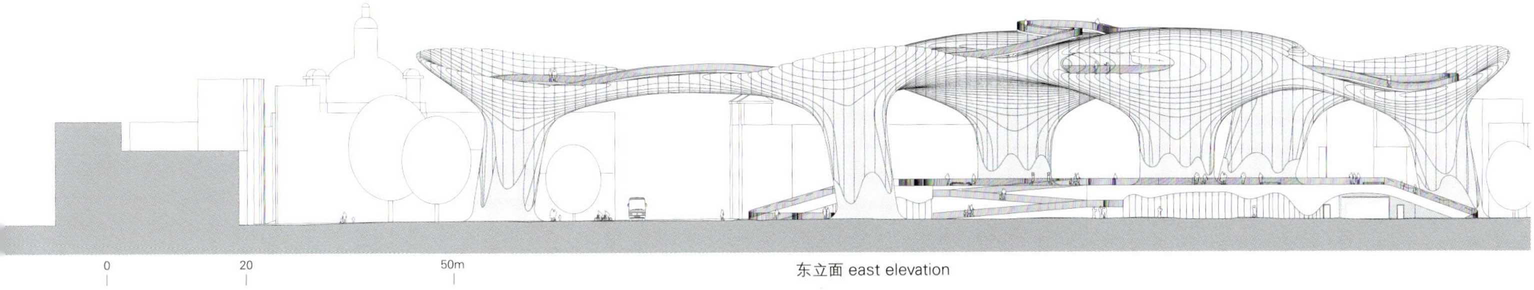

东立面 east elevation

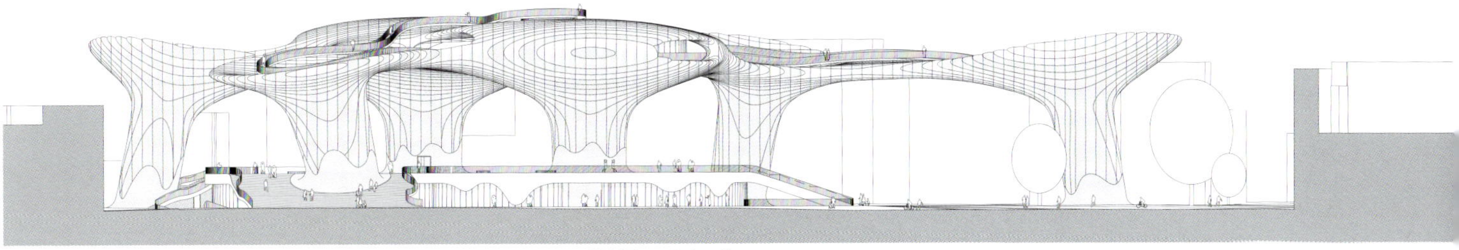

西立面 west elevation

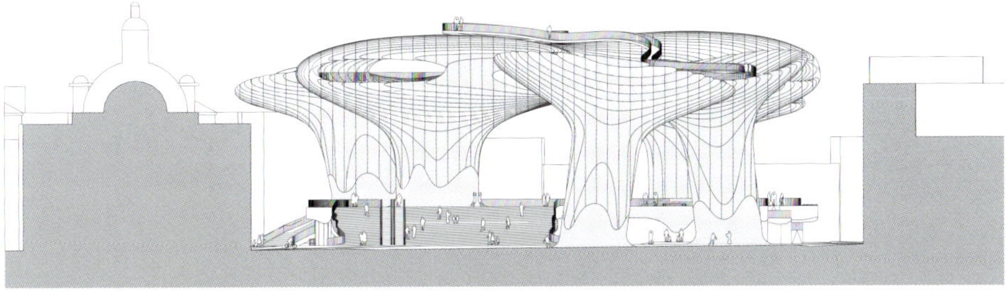

南立面 south elevation

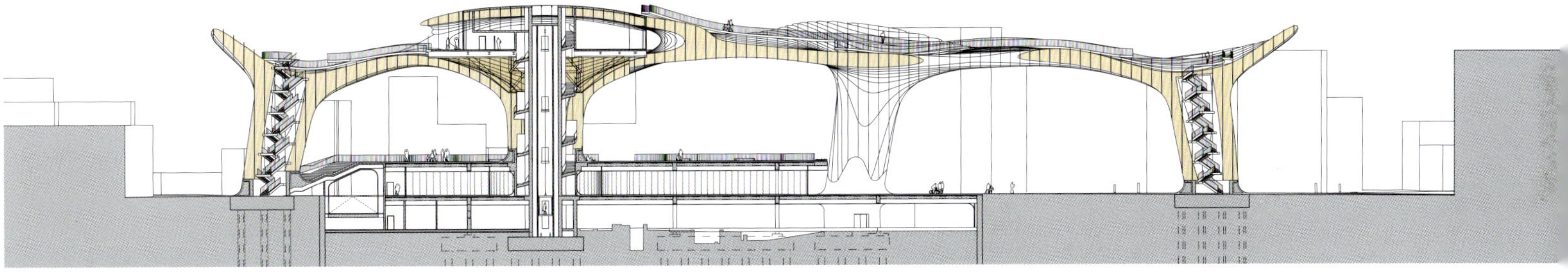

A–A' 剖面图 section A-A'

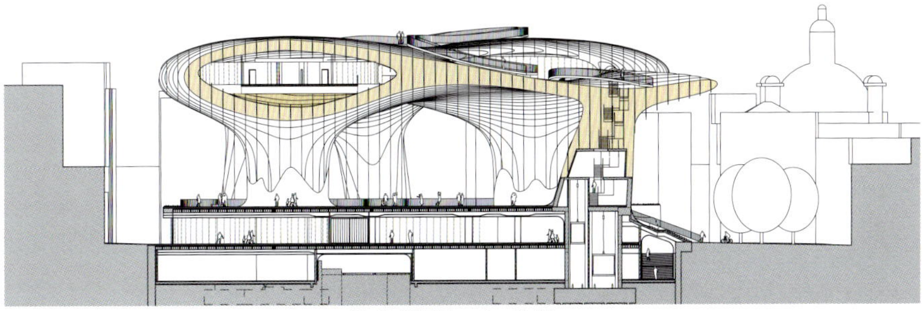

B–B' 剖面图 section B-B'

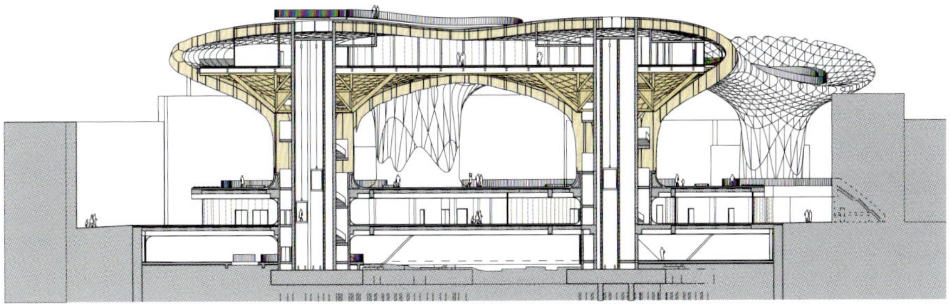

C–C' 剖面图 section C-C'

HOTEL DUCAL

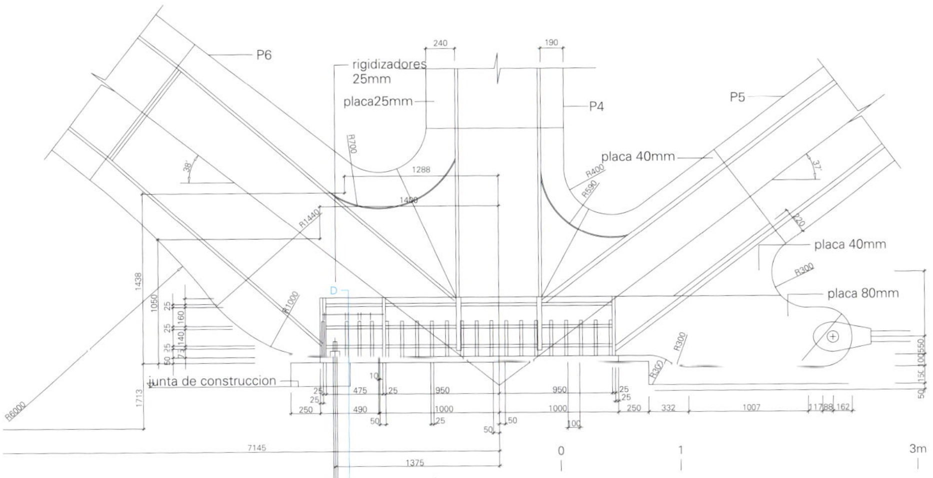

五根支柱 V vista support pillar

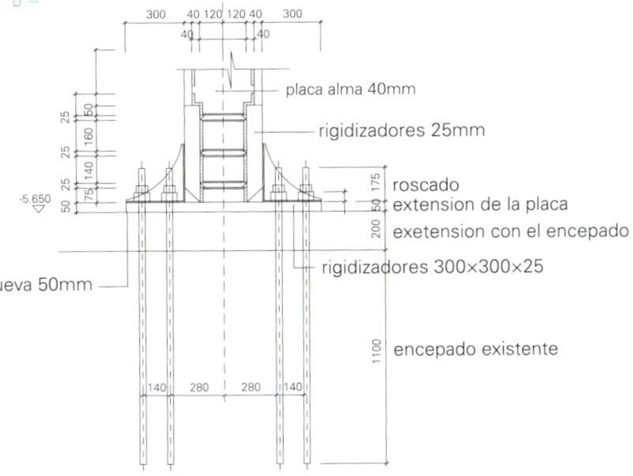

五根支柱 D–D' 剖面图 V vista support pillar section D-D'

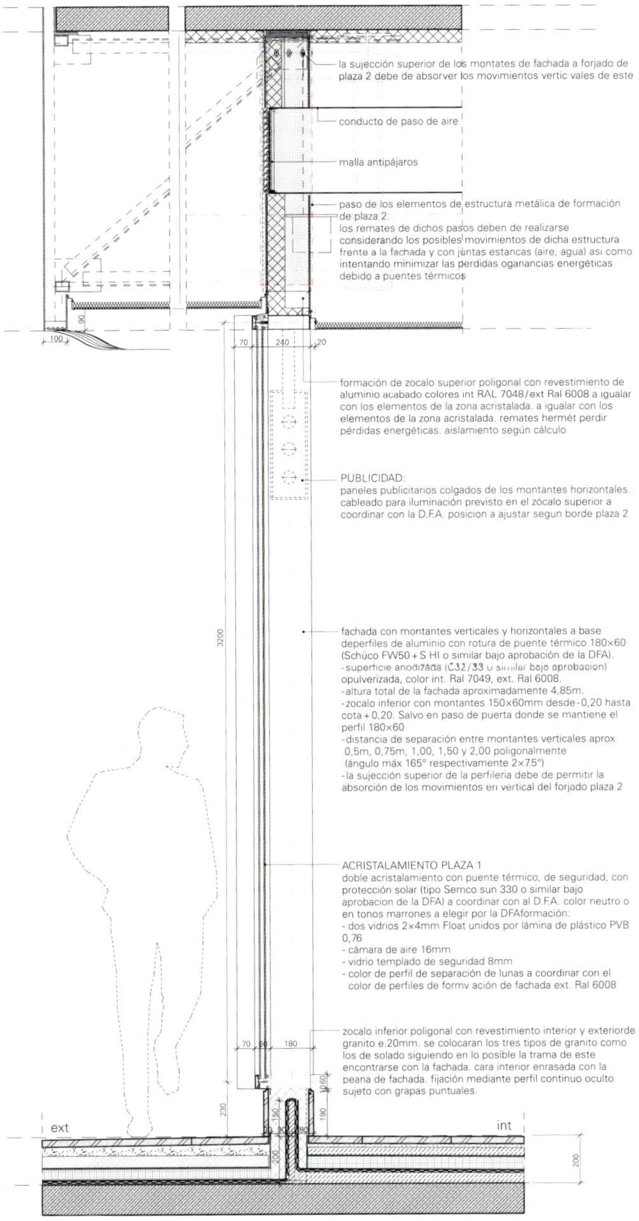

玻璃面和广告牌 glazing and billboard

precious shade during the day, creating a comfortable micro-climate and a place for relaxation. At night, the MP becomes an artificial sky that sets the stage for various light and sound scenarios. These two qualities allow for a series of urban activities such as sports (beach volleyball, boxing, street basketball, etc.), cultural events (cinema, theater, concerts, etc.) as well as commercial uses (car-and fashion shows, corporate events, presentations, etc.) that emphasize the Plaza's role as one of the city's main places for communication and interaction.

The Elevated Plaza responds to the need for a large, multifunctional urban meeting space in the old city center of Seville. It is connected to the street level via large stairways that function as activators for street life. While being large enough to mark the plaza as a public space, the stairways make for easy reconfiguration in the event of restricted access such as concerts, theater performances or fashion shows. Holes in the platform visually connect it to the market underneath and further to the museum underground. The architectural proposal doubles the original area and creates an additional 4,500m² of high quality programmable space. Water basins located in selected areas of the surface further enhance the micro-climate and offer attractive locations for sitting and lingering.

After some months of researching structural possibilities we decided to use the Mensa Project as the prototype. Metropol Parasol will also be built with laminated wood plates and covered with a polyurethane skin. Of course the lower levels will be steel and concrete, and the areas of the restaurant in the parasol have concrete shafts and steel substructure because of fire regulations. But, the parasols themselves are wood construction. We considered very complex parameters like prefabrication, maintenance, costs, life span, fire loads, earthquake and traffic loads, temperature reactions to name a few. Wood always survived as the best solution.

Metropol Parasol redefines the potential of Plaza de la Encarnación to become the contemporary urban center of Seville. The architecture proposed here will thus not only create a new place of identification for the people of Seville but also articulate Seville's role as one of Spain's most fascinating destinations.

J. Mayer H. Architects

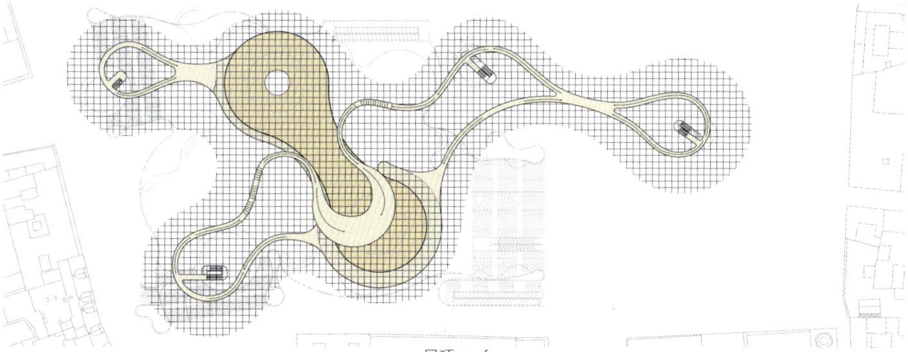

屋顶 roof

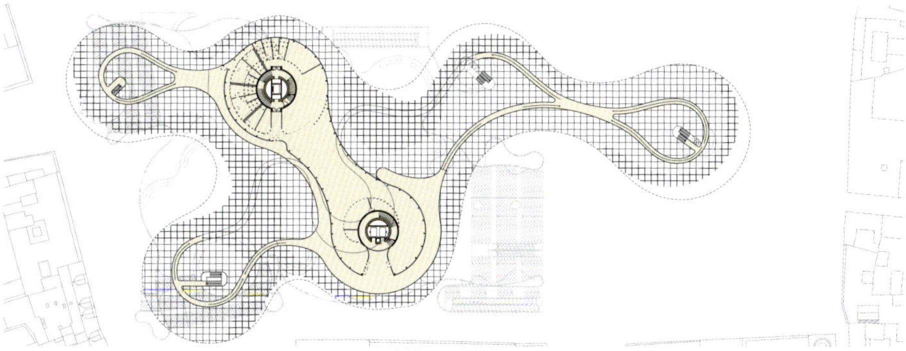

七层 seventh floor

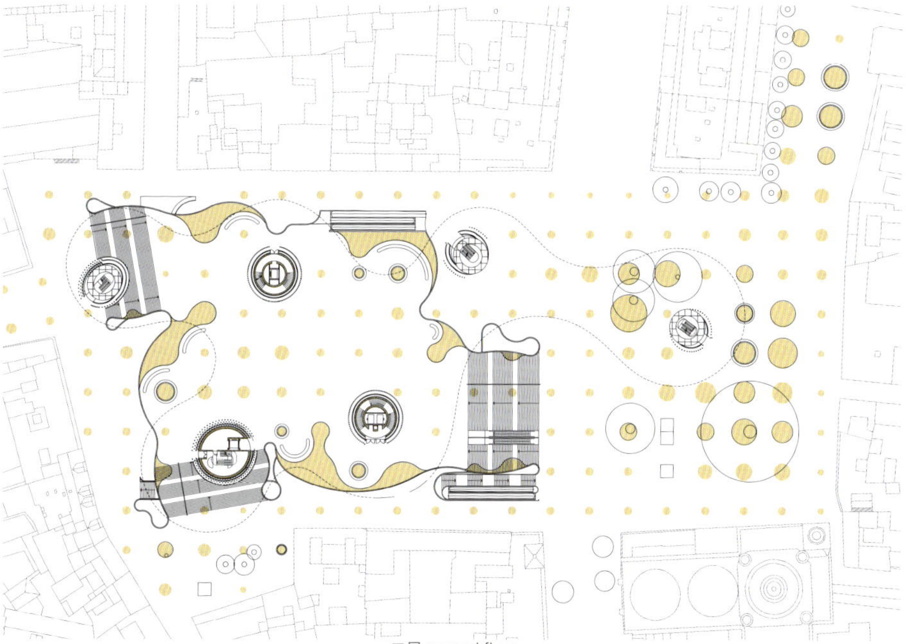

二层 second floor

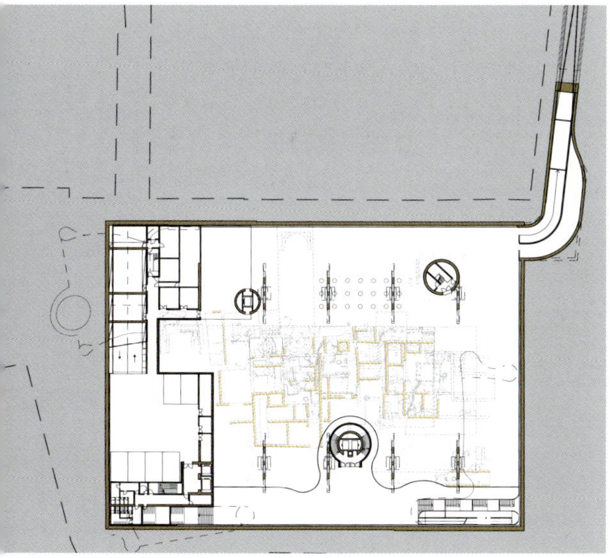

地下一层 first floor below ground

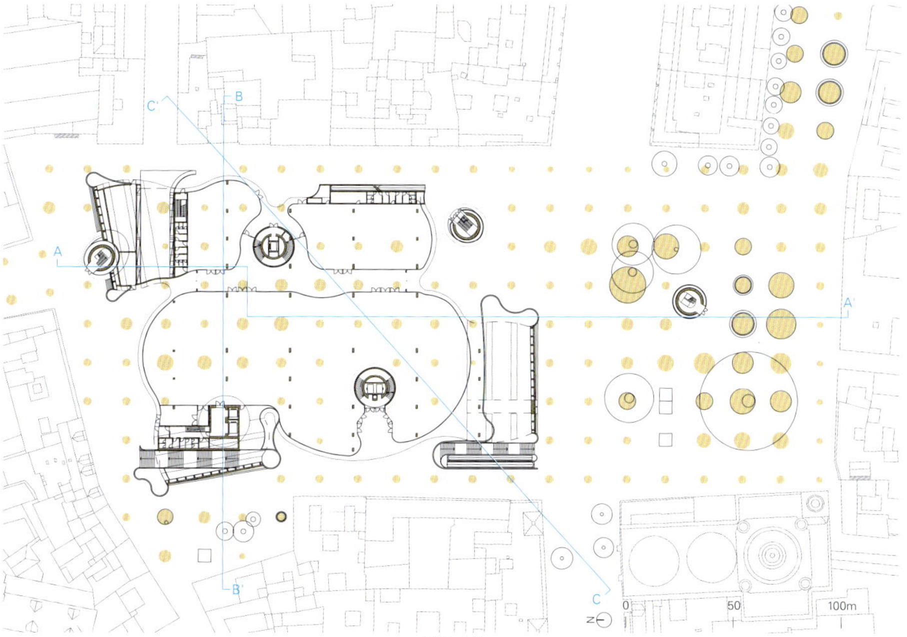

一层 first floor

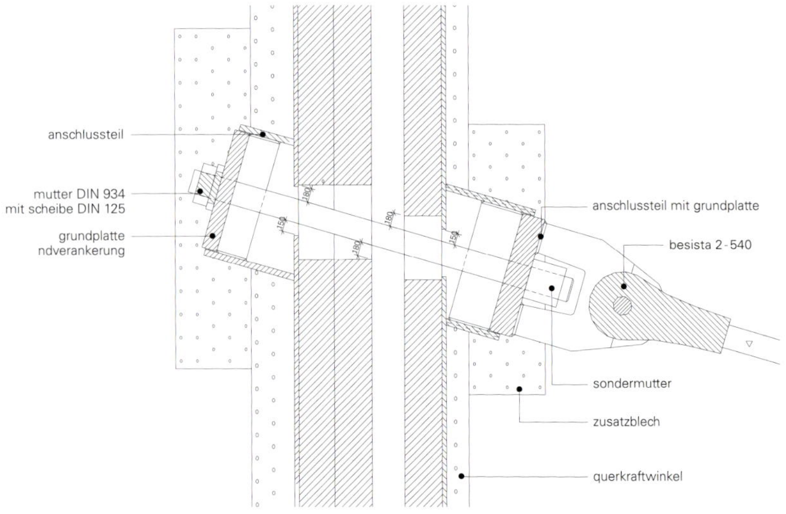

剪力张拉杆托架 shear force tension rod bracket

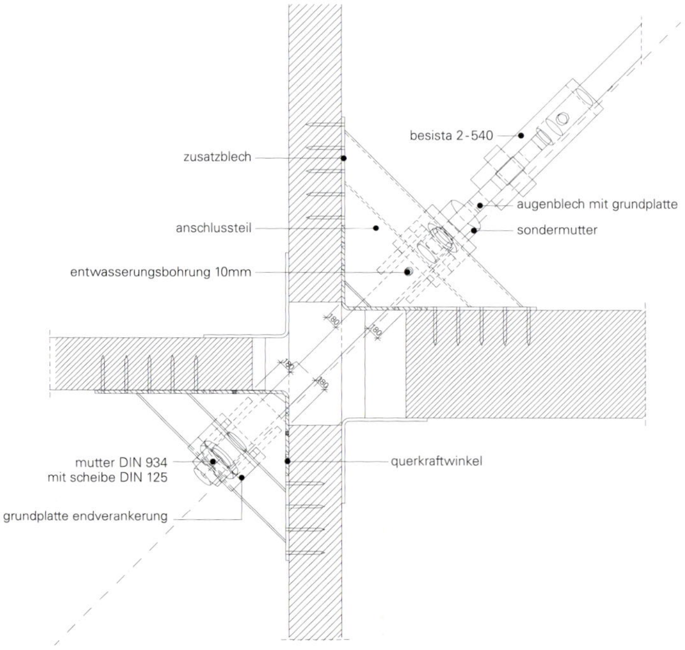

详图1 detail 1

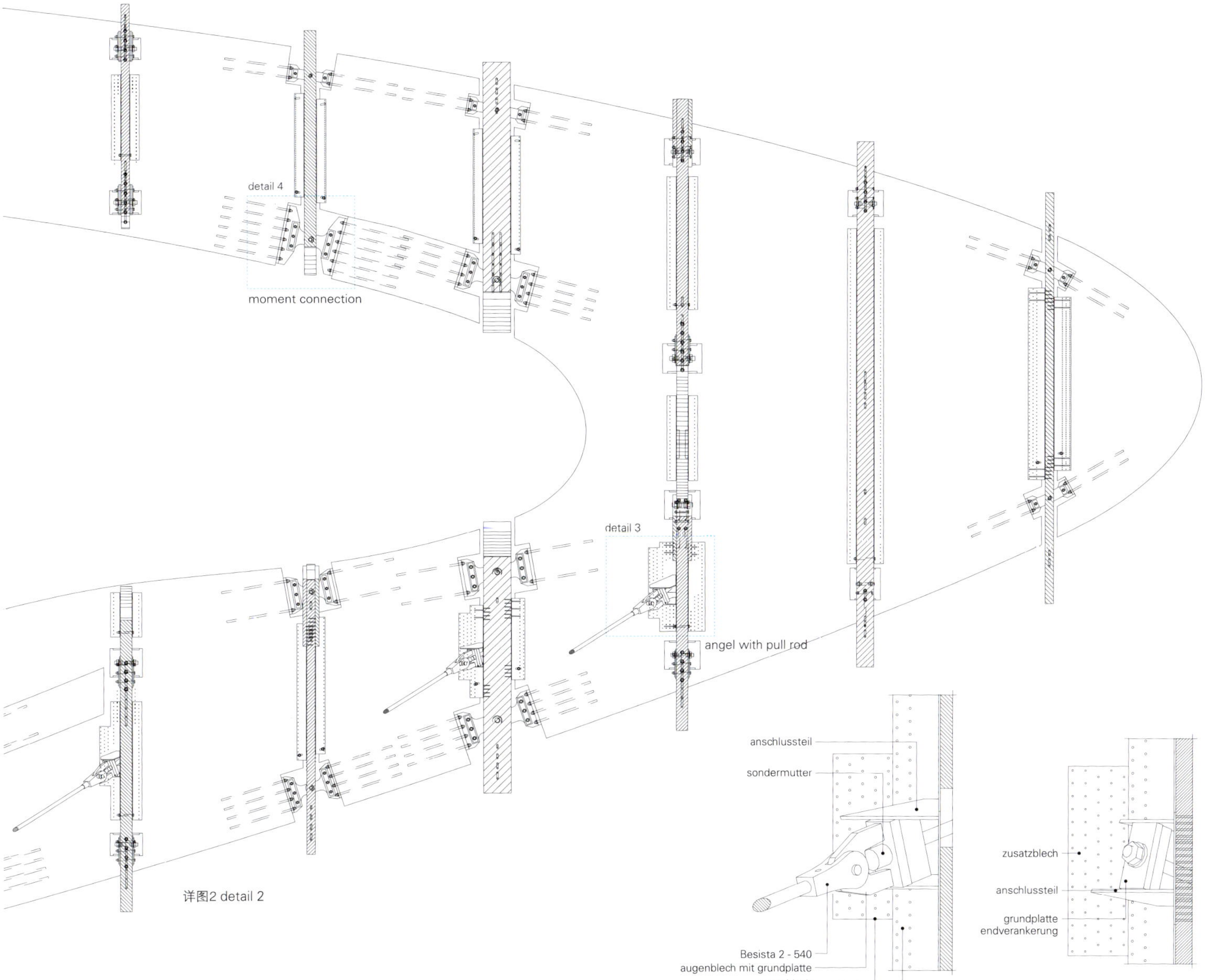

详图2 detail 2

详图3 detail 3

courtesy of the architect (©Ignacio Ysasi)

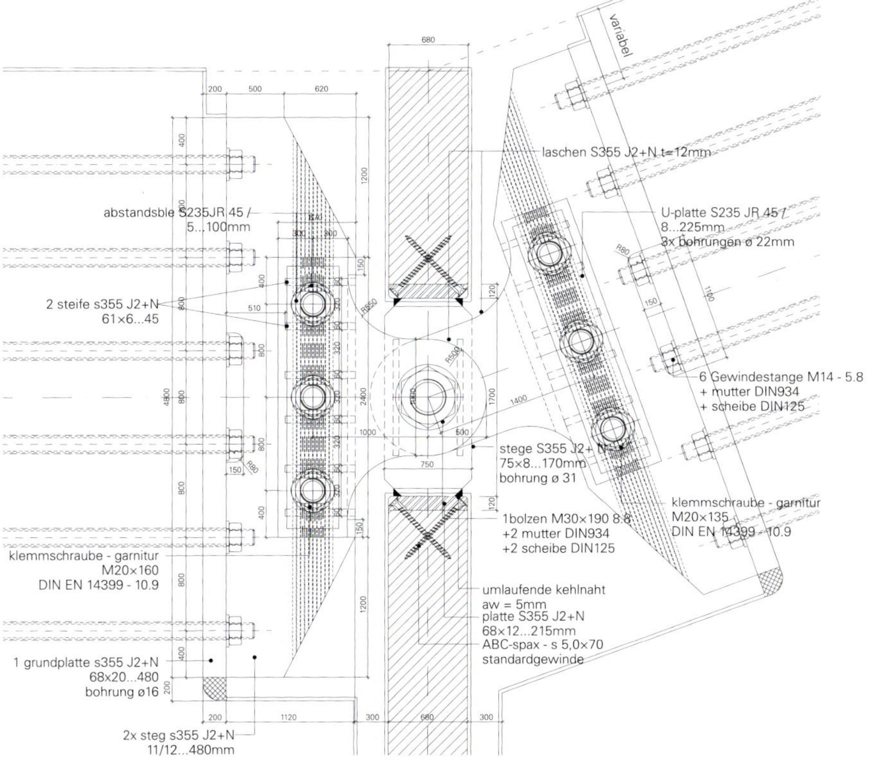

详图4 detail 4

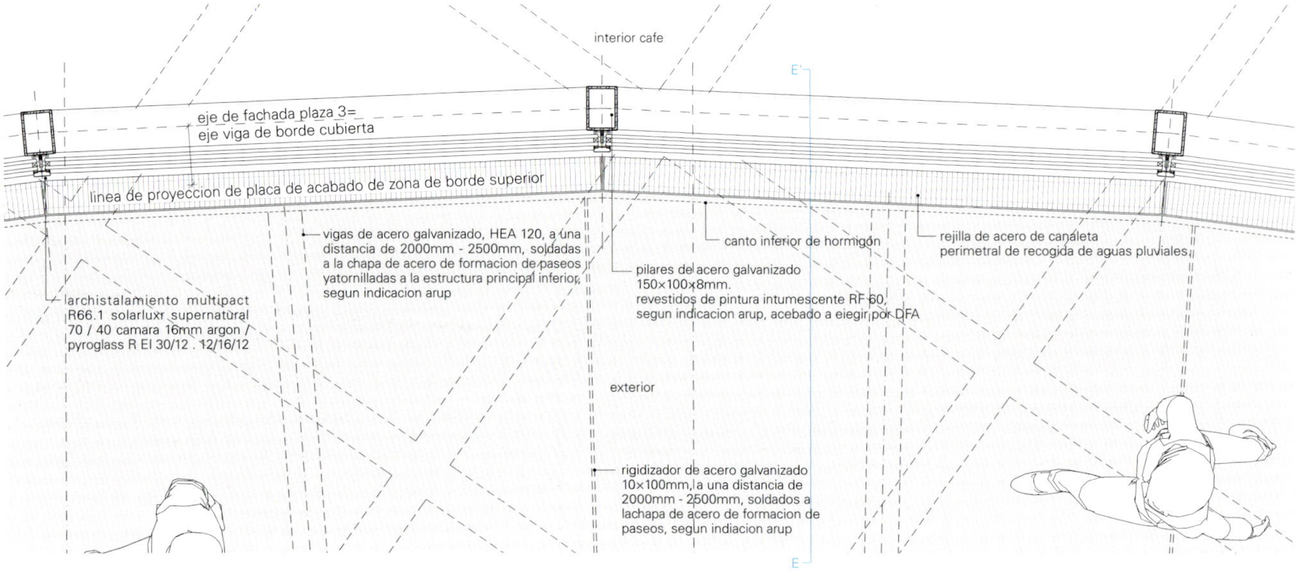
interior cafe
eje de fachada plaza 3=
eje viga de borde cubierta
linea de proyeccion de placa de acabado de zona de borde superior
vigas de acero galvanizado, HEA 120, a una distancia de 2000mm - 2500mm, soldadas a la chapa de acero de formacion de paseos yatornilladas a la estructura principal inferior, segun indicacion arup
canto inferior de hormigon
rejilla de acero de canaleta perimetral de recogida de aguas pluviales
larchistalamiento multipact R66.1 solarluxr supernatural 70 / 40 camara 16mm argon / pyroglass R EI 30/12 . 12/16/12
pilares del acero galvanizado 150×100×8mm.
revestidos de pintura intumescente RF 60, segun indicacion arup, acebado a elegir por DFA
exterior
rigidizador de acero galvanizado 10×100mm, a una distancia de 2000mm - 2500mm, soldados a lachapa de acero de formacion de paseos, segun indiacion arup
E'
E

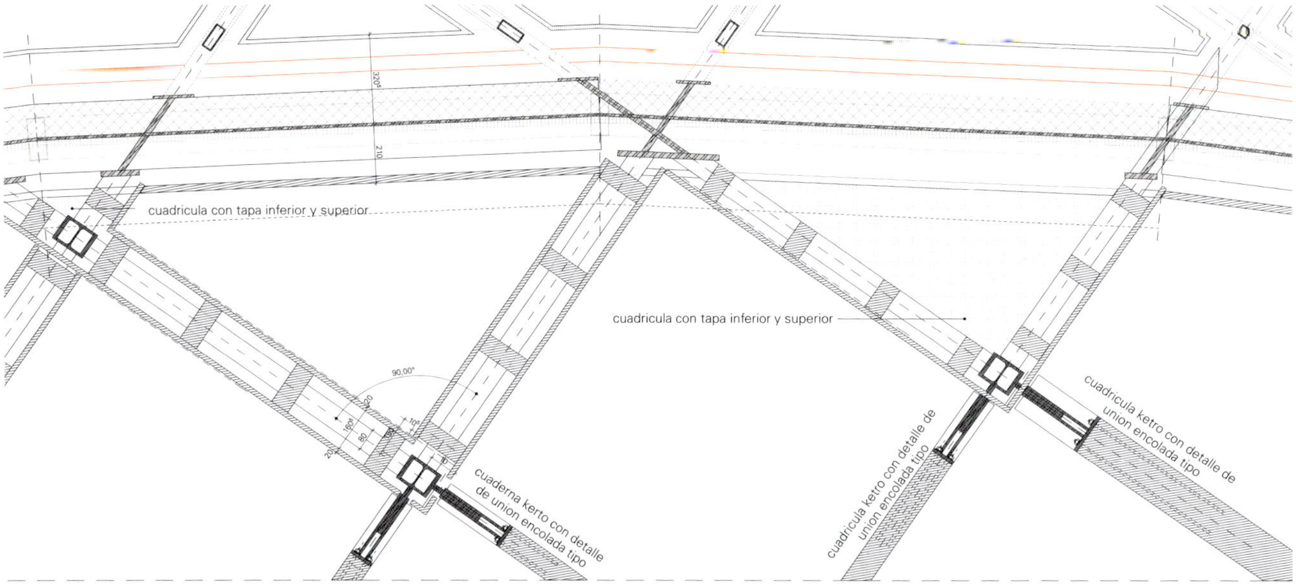
cuadricula con tapa inferior y superior
cuadricula con tapa inferior y superior
90,00°
cuaderna kerto con detalle de union encolada tipo
cuadricula ketro con detalle de union encolada tipo
cuadricula ketro con detalle de union encolada tipo

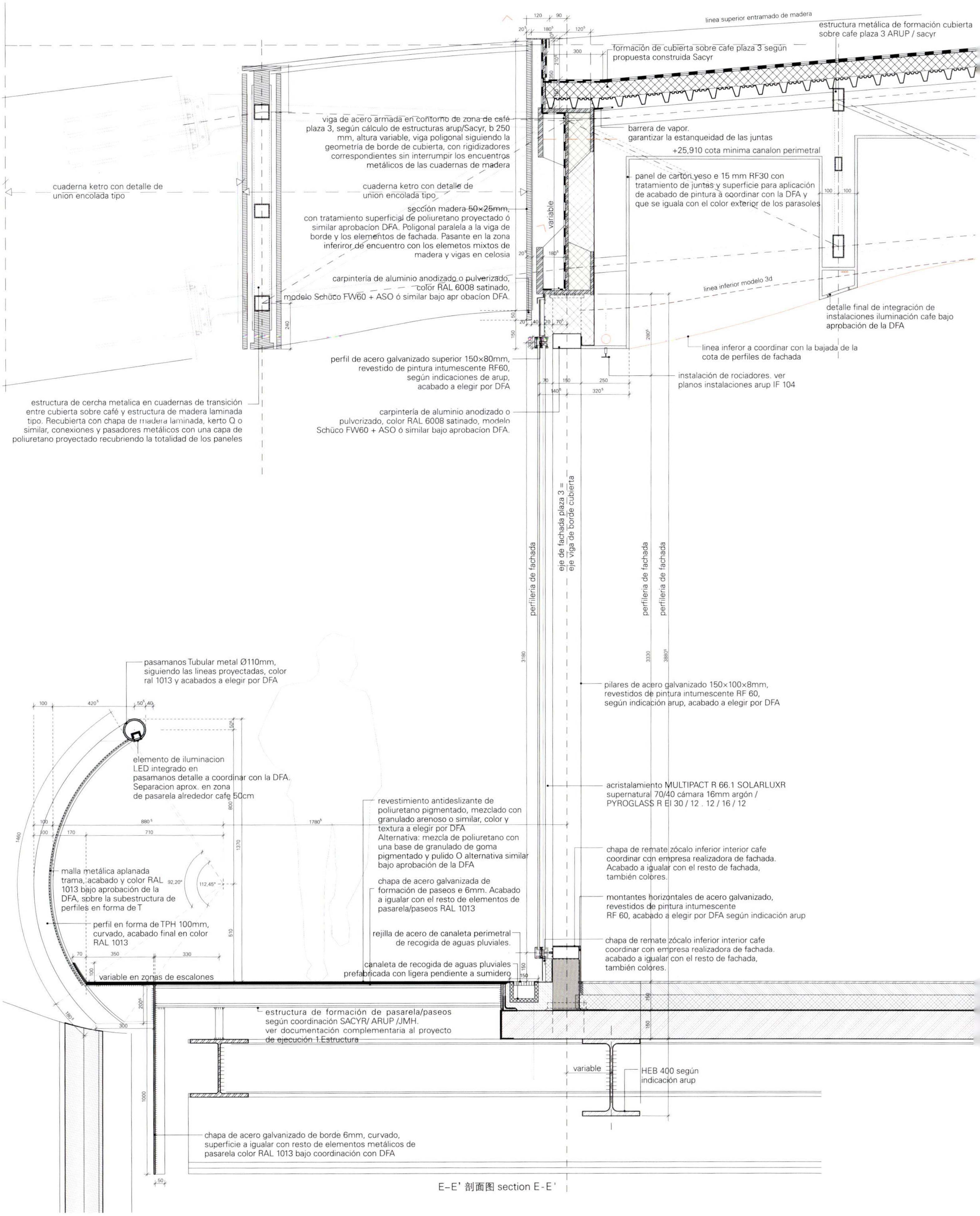

E–E' 剖面图 section E-E'

"走，慢慢走，用心演绎"是皮耶罗•钱皮演唱的一首歌曲，描绘了一系列主要利用通道吸引参观者的博物馆。在博物馆中漫步相当于通过身体移动来实现艺术演绎，该情景正如罗伯特•瓦尔泽在小说《散步》中所描绘的那样。

透过钱皮所用的动词，我们对当代博物馆做了一次调查，就像布莱恩•奥多尔蒂用文字对建筑进行评估一样。在钱皮吟唱那首歌曲的岁月里，布莱恩•奥多尔蒂对一间变成白色立方体的博物馆进行了深入的评鉴。走过室内路径之后，我们自然会过渡到室外，过渡到博物馆自身就是艺术品的层面。按照爱尔兰艺术家z的说法，这间博物馆是白立方之子。

最后，格哈德•里希特将向我们展示一种改变封闭性博物馆的方法，即拥抱美景，以"走，慢慢走"的方式开始此次旅程——正如意大利歌手演唱的那样——最后穿过景观入口作为结束。

Going, walking, working is a song by Piero Ciampi that introduces the description of a group of museums that use the path as a main attraction. Walking in a museum means performing an artistic act through a physical movement, such as Robert Walser describes in the novel *The Walk*.

Through the verbs proposed by Ciampi, we carry out a survey on these contemporary museums, as assessed in the light of the words of Brian O'Doherty, who, in the same years when Ciampi was singing his song, presented a profound criticism of the museum turned into a white cube. From walking in interior spaces, we therefore will pass to the outside, to the concept of the museum as a work of art in itself, son of the white cube described by the Irish artist.z

Finally, Gerhard Richter will show us a way to undermine the museum box, through the opening to the landscape, concluding the journey started with the verbs going, walking, by the Italian singer.

走，慢慢走，拥抱美景
听一首歌，听书中讲述的博物馆故事

Going, Walking, Opening to the Landscape Museum Stories told by Books, listening to a Song

走，慢慢走，拥抱美景
听一首歌，听书中讲述的博物馆故事

走，慢慢走：关于艺术，运动中的艺术

1975年，意大利歌手和歌曲作家皮耶罗•钱皮写了一首讽刺性的——几乎是有预见性的——劝谏歌曲来警醒世人。20世纪80年代后的世界将会见证一股类似强迫性乐观主义潮流和当代放纵派吟游诗人对之前数十年间（从第二次世界大战到1968年的示威游行）发生的重大事件的遗忘。这股潮流似乎塑造出了道德水准远高于20世纪70年代末的一代人，那一代人对消费主义有着近乎狂野的崇拜。

这种讽刺现象在人类行为上有着集中而深刻的反映："走，慢慢走"，这些动词可以被看作是人类返璞归真的实例，就好似人们用自己身体完成的旅程构成了人类真正的本性、人类灵魂的主要特征。然而，钱皮将会搜集他歌中所唱的那些辛辣语句，将自己变成一个如飞蛾扑火般走向自我毁灭的流浪汉，在旅行的过程中，通过与截然不同的人接触探索出一个全新的交友圈，这是他丰富歌曲艺术及文化涵养的一条基本原则。

散步变成了一种纯粹的思想行为和艺术行为，毫不意外地成为一本书的主题词，而这本于1919年由罗伯特•瓦尔泽撰写的书——《散步》，早在近六十年前就预见了钱皮专辑的诞生。

"我必须要告诉大家，在一个美好的早晨，我不确定是几点几分，心里突然冒出去散步的念头，我把帽子戴在头上，离开我的写字间——那幽灵游荡的房间，跑下楼梯，冲上大街……

我记得，当我站在阳光明媚的街道上，我处在一种浪漫、刺激的情绪中，这让我很开心。摆在我面前的晨光中的世界看起来是如此的美丽，仿佛那才是我们的初见。我眼中的一切都给了我一种愉悦的感受——亲昵、和善、青春洋溢。简言之，我竟然忘了就在最近，就在我楼上的房间里，我还独自一人浑浑噩噩地躺在白床单上做着白日梦。虽然我仍然能强烈地感受到在我身前身后都弥漫着一些肃穆的音符，但悲伤、痛苦和所有阴暗的想法都不见了。

这让我对现在所经历或未来将要经历的种种都充满了美好的憧憬。我的步伐平稳且镇定，我想在我这样行走的过程中已经显示出一种相当优雅的风度。"

——罗伯特•瓦尔泽，《散步》，1919年

在这部短篇小说中，我们鉴赏了这位意大利歌手所使用的动词——"走，慢慢走"，通过沿途一连串的事件，与人、物、景观、建筑不时的偶遇得以体现。故事以一种绵长、意识流式的结构展开，很像乔伊斯所著《尤利西斯》中的主人公利奥波德•布鲁姆所走的接近现代的道路。

Going, walking, opening to the landscape
Museum stories told by books, listening to a song

Going, walking: about art, in moving

In 1975, Piero Ciampi, Italian singer and songwriter, wrote an ironic exhortation, almost predictive, to a world that, from the eighties, will see the emergence of a kind of forced optimism and a contemporary goliardic oblivion against major events of the previous decades (from the Second World War to the 1968 Protests), which seemed to have forged a generation of people with an ethic far superior to that shown in the frenzied consumerism that one will witness from the end of the seventies.

This irony has a profound reflection focused on the action of man: "going, walking", verbs seen as possible examples of a return to the essence of life, as if the journey, especially the one done with one's body, constitutes the true nature of people, the main feature of the human soul. However, Ciampi will collect the challenge proposed in the words of his song, becoming a tramp who, with a fury close to the limit of self-destruction, searches in his travels a new sociality in meeting totally different people, as a basic principle of its artistic and cultural enrichment.

Walking becomes a pure act of thought, an artistic act, not surprisingly the protagonist of a book that anticipates by nearly sixty years the album of Ciampi: *The Walk*, written by Robert Walser in 1919.

I have to report that one fine morning, I do not know for sure what time it was, as the desire to take a walk came over me, I put my hat on my head, left my writing room, or room of phantoms, and ran down the stairs to hurry out into the street. [...]

As far as I can remember, just as I was on the sunny street I was in a romantic and adventurous disposition, which made me happy. The morning world that lay before me looked to me so beautiful as if I was seeing it for the first time. All I could see gave me a pleasant feeling of affection, kindness, youth. In short, I forgot that until recently, up in my room, I was left to daydream gloomily on a white sheet. Sadness, pain, and all the dark thoughts were missing, although I continued to perceive accurately, before and behind me, some serious notes.

It filled me with joyful expectation of all that could meet me or present to me. My steps were measured and calm, and I think I've shown, while so I walked, a quite decent demeanor.

– Robert Walser, *The Walk*, 1919.

In this short novel, we watch the description of verbs proposed by the Italian singer, "going, walking", in the form of continuous accidents along the path, accidents from time to time made by encounters with people, objects, landscapes, and architectures. The paper proceeds as a long and structured stream of consciousness, much like the nearly contemporary path taken by Leopold Bloom in Joyce's *Ulysses*.

The life and therefore the art, which is nothing but a discussion on life itself, is enhanced in the movement of humans, through contacts between the senses of the same ones. The writing of a novel, as a song, can simply build itself through the description of a movement: "going, walking" do not need to be explained; it is

走，慢慢走，用心演绎
昂首阔步，
胆小、麻木的人，
愧疚、绝望的人。
不要沮丧，我们走，
去演绎，
走，慢慢走，用心演绎，
美酒对石油，
赢得漂亮，赢得精彩，
赢得畅快淋漓。
走，慢慢走，用心演绎，
南边咆哮，
北边还没上涨，
别担心，这里有下跌，
走，慢慢走，用心演绎，
快点逃，快点逃，
逃快点，逃快点。
走，慢慢走，用心演绎，
地痞都关进了监狱
恶犬都关进了狗舍
阳台上有玫瑰
庭院里有猫咪……
走，慢慢走，用心演绎
走，慢慢走，用心演绎
来吧，去演绎！
——皮耶罗•钱皮，“走，慢慢走，用心演绎，走，慢慢走，用心演绎……及其他演讲”，1975年

生活以及源于生活的艺术——艺术就是对生活的探讨——在人类的运动过程中得到了加强，也通过与人类的感官沟通得以加强。一本小说的创作，就如同一首歌曲的创作，可以通过单纯描述一个运动的过程，如“走，慢慢走”来实现，无需过多的解释，本身就已足够。事实上，这就是对行为的描述。

走，慢慢走：内部，博物馆内部

“走，慢慢走”，这两个动词与艺术思想之间的联系非常紧密，可以在下文介绍的博物馆空间发展过程中得到体现，这些博物馆建筑共六个，包括：由Odile Decq Benoît Cornette Architectes Urbanistes设计建造的MACRO当代艺术博物馆、由FREE-Fernando Romero设计建造的索玛雅博物馆、由Nieto Sobejano Arquitectos设计建造的圣特尔莫博物馆的扩建、由Rocco Design Architects建造的广东省博物馆、由Alberto Campo Baeza设计建造的安达卢西亚历史博物馆和由Camilo Rebelo + Tiago Pimentel设计建造的Côa谷艺术考古博物馆。

在上述所有建筑中，设计者都特别关注游人在建筑内部的各个空间中穿行的流线设计。它们不仅仅是展览展品的通道，而且都或多或少地因馆内装置的不同而各具特色，这些装置有些是临时性的，有些则是永久性的；这些建筑好似围绕着各种机械装置而建，而这些装置指明了参观者通道：楼梯、坡道、悬挑走道、细栏杆和窄桥。

所有博物馆都利用空间装置进行运作，这些装置试图让通道具

Going walking working,
going dearly,
band of timids, of unconsciouses,
of indebteds and desperates.
No dejections, let's go,
go to work,
going walking working,
the wine against the oil,
big win, big win,
great victory.
Going walking working,
the South roars,
the North has not climbs,
do not worry, here is the descent,
going walking working,
quick getaways quick,
getaways quick getaways.
Going walking working,
bullies all locked
dogs with the dogs in kennels
roses on the balconies
cats in the courts…
Going walking working
Going walking working
Come on, work!
– Piero Ciampi, *Going walking working, Going walking working… And Other Speeches,* 1975

enough, in fact, the description of the action.

Going, walking: inside, in the museum

"Going, walking", verbs so intimately tied to an artistic thought, easily find themselves in the spatial development of the museums presented below: the MACRO Museum of Contemporary Art by Odile Decq Benoît Cornette Architectes Urbanistes, the Soumaya Museum by FREE-Fernando Romero, the San Telmo Museum Extension by Nieto Sobejano Arquitectos, the Guangdong Museum by Rocco Design Architects, the Andalucia's Museum of Memory by Alberto Campo Baeza, and the Côa Valley Museum of Art and Archaeology by Camilo Rebelo+Tiago Pimentel.

In all the buildings, a special emphasis is placed in the act of crossing the spaces that compose the buildings. They are not simply exhibition paths, more or less characterized by the installations, temporary or permanent, of the works; the buildings seem to be built around mechanisms that show the path of visitors: stairs, ramps, suspended walkways, slender parapets, thin bridges.

All museums act with spatial devices that tend to implement the path with continuous and amazing sensations, as to recreate some landscapes, now unknown in the cities. So, the interior spaces are transformed into places of active thinking, a kind of concentrated training of what could be meetings, accidents, landscapes, lands covered during long walks. The exhibited artworks, therefore, become functional to paths themselves, not vice versa: they become part of those meetings that Walser described in his novel, as the roses on the balcony sung by Ciampi.

画廊通道，MACRO博物馆，罗马
gallery passage, MACRO, Rome

备持续、迷人的感官特征，好似重塑某些如今在城市里已经无人知晓的景观。由此，室内空间变成了灵感迸发的场所，就像一个集训过程，让人们在悠长的漫步过程中了解何为会面、何为偶遇、何为景观、何为土地。因此，展出的艺术品对通道本身来说具有一定的作用，但反过来却并非如此：它们成为瓦尔泽小说中所描绘的会面场所的一部分，就像钱皮歌中唱的“阳台上的玫瑰”一样。

在一些展示的项目中，这种想法被迫变成了一种独立的组成元素，整座建筑都围绕着它修建。

安达卢西亚历史博物馆的坡道好似一个静置的、纹丝不动的棺椁外壳上跳动的心脏，是整个无懈可击的建筑外壳上的一个突破口。坡道穿透屋顶，开始与外界和天空进行互动；拾级而上的人流站在巨大圆形空间的中央，在纯白色背景的映衬下凸显出来，成为一片静态环境中丰富的动态因素。

然而，MACRO博物馆的心脏是红色的，即使在欧几里得几何体系中也找不到绝对完美的诠释：这间博物馆强调的是无序性；从场馆中央、礼堂到轻质走道分支。悬吊的场所增加了空间的密度，可供人们驻足，欣赏独一无二的景色。

在广东省博物馆中，一系列从盖板上悬垂下来的楼板使中央空间变得灵动，让空间有了形状，否则，室内空间虽然很有个性，但却是触不到的，缺乏真实感。设计者建造了一处类似中央庭院的景观，是参观者不能自由穿行的一片区域，充足的日光从天窗洒落，景观中央只有一条窄路可供通行，引领参观者体会置身真空之中的惊险与刺激。

圣特尔莫博物馆拒绝效仿之前博物馆的恢弘气势，而选择修建令人感觉更亲切的道路，引领人们获得更澄净透彻的感悟。长门厅颀长的体量似乎吸引着参观者走到尽头，一路上交替出现开放式空间和露台，使人们在经过展厅的通道之后重回自然与宁静。

因此，进入弯曲的空间会使人们的思绪更加混乱。参观者们实际是步入一片纯白色的区域，虽然素净，但极其鲜活、纷杂，近乎猛烈。这就是索玛雅博物馆内的空间，馆内的道路设计对行走其上的参观者永远是一种挑战。这里有不计其数的可能性，如此复杂的内部结构才能与外部的个性——拉长、扭曲、不稳定的平衡感相媲美。

Côa谷艺术考古博物馆在室外景观里找到了整个项目的关键性要素：一条路像蛇一般没入地下，蜿蜒起伏，最终成为博物馆内部建设的一份奖赏。博物馆作为土壤固化的产物，碎裂成一系列立面，是人们尽情欣赏大扇窗户框起来的全景的最后地方。

走，慢慢走：奇观，在室外

“现代主义的历史与（画廊）空间密不可分；更确切地说，现代艺术的历史与那个空间内发生的变化以及我们看待它的方式息息相关。我们现在已经达到一个‘不见艺术，先见空间’的阶段。（这个时代老套的做法是一进画廊，先就空间进行讨论）。一片白色、唯美的空间——而不是哪一幅画作——或许是20世纪艺术的典型形象；这一时期的艺术通过一段历史必经的过程证明自己，而这种过程往往是与

In some of the presented projects, this attitude is forced to the point of becoming a single compositive element, around which the whole building is constructed.

The ramp of the Andalucia's Museum of Memory seems the pulsating heart of an exterior casket, motionless and imperturbable, a breaking point in the perfection of architectural container. The ramp breaks through the roof and starts interacting with the outside, with the sky; streams of people, then being stepped up, stand out in the pure white of the large circular hole, as rich variables in a static composition.

The heart of MACRO, however, is red and cannot find absolute perfection in Euclidean geometry: here the disorder is emphasized; from the center of the museum, the auditorium, light walkways branch and densify the space with flying plots, on which one can observe original views.

In the Guangdong Museum, a series of floors hanging from the cover animate a central vacuum, giving shape to a volumetric form, otherwise intangible but highly characterizing the interior space. As an impenetrable area, a sort of central courtyard is defined, marked by the light that rains abundantly from the huge skylight and crossed only by a narrow walkway, which leads the visitor to experience the thrill of the vacuum.

The San Telmo Museum rejects the spectacularity of the previous examples, building more intimate paths that lead to a more sober reflection. Long halls, where the stretched size seems to attract the visitors toward the bottom, alternate with open spaces and patios that reconstruct a natural calm after the exhibition path.

Thus, aspirated within warped spaces, but with a much more chaotic attitude, visitors walk into an area of pure white, and yet tremendously lively, messy, almost violent. They are the spaces enclosed in the Soumaya Museum, in which the paths are set as a constant challenge to people who walk over them. The possibilities are numerous, superabundant, as to create a complexity that can withstand the personality of the outside – stretched, distorted, in unstable equilibrium.

The Côa Valley Museum of Art and Archaeology finds in the outside landscape the key of the overall project: as a serpent; one road unravels itself into the earth, winding and bumpy, and ends up as a prize to the physical effort in the museum which, as a concretion of the soil, breaks into a series of facets, a last difficult setting to fully enjoy the final panorama framed by large windows.

Going, walking: the astonishment, in the outside

The history of modernism is intimately framed by (the gallery) space; or rather the history of modern art can be correlated with changes in that space and in the way we see it. We have now reached a point where we see not the art but the space first. (A clichè of the age is to ejaculate over the space on entering a gallery). An image comes to mind of a white, ideal space that, more than any single picture, may be the archetypal image of twentieth century art; it clarifies itself through a process of historical inevitability usually attached to the art it contains.

The ideal gallery subtracts from the artwork all cues that interfere with

古根海姆博物馆的圆形大厅，纽约，1959年，弗兰克•劳埃德•赖特，克里斯•伊森提供
rotunda of Guggenheim Museum, New York, 1959, Frank Lloyd Wright ©Chris Eason

它所包含的艺术相互依存的。”

“理想的画廊能从艺术品中摒弃一切妨碍它成为‘艺术品’的因素。作品远离任何有可能干扰其发展自身的因素。由此，这间画廊与其他保留传统方法——不断重复封闭价值体系——的画廊一样，在艺术界拥有了一席之地。教堂的神圣、法庭的庄重、实验室的神秘与设计的时尚融合在一起，形成了一个独特的美学空间。这个空间内的感知力场极其强大，所以艺术一旦到这个空间之外就会沦落为粗俗的玩意儿。相反，当置身于一片艺术理念强盛的空间之中，艺术的光芒笼罩其上时，普通的物品也能变成艺术品。”

——“在白立方内”。“画廊空间的意识形态”，布莱恩•奥多尔蒂，《工作室与立方体》，1999年（1976年在《艺术论坛》杂志上发表的若干篇文章）

1976年，在《艺术论坛》杂志中，爱尔兰艺术家布莱恩•奥多尔蒂发出对展览空间的演变过程做一次反思的倡议，尤其是当素净的白色展厅成为并持续作为艺术品展主要空间背景之后的那段历史。根据这位艺术家的说法，我们经历过用简单的画框将一幅作品与另一幅作品分隔开来的阶段，而后过渡到将同一幅作品安放在一片尽可能无色彩的空间中，从而在其周围形成一片气体真空，将参观者的注意力完全集中到作品上。

这一过程改变了建筑的面貌，赋予悬挂展品的墙壁以非凡的重要性，使它脱离艺术品，使博物馆不再臣服于艺术家，即使不比它们更重要，也至少是与其平起平坐的。如果这种反思在1976年是正确的，尤其是对画廊——更确切地说，是对博物馆内部——而言是正确的，那么时至今日博物馆建筑在功能方面的形变则已经拓展到了外壳，即建筑的外观上。那两座最为著名的建筑——分别位于纽约和毕尔巴鄂的古根海姆博物馆——很好地诠释了这一演化过程；如果说赖特设计的展厅通道迫使驻馆艺术家们挑战那片由一条绵延的坡道包围而成的空间，那么盖里设计的博物馆则对展览这一主要功能完全没有兴趣。这间西班牙博物馆是为其他用途设计的：整座建筑化身成为新城区规划的插曲，而新城区规划正是该市城区发展的焦点之一。为了实现这一目标，盖里将艺术范围无限扩大，不考虑某一具体地点，而是将城市作为一个整体考虑。为了修建一条丰富多彩的通道——“走，慢慢走”——他需要将博物馆本身转化成一件艺术品。位于毕尔巴鄂的古根海姆博物馆则是一次都市偶遇的结果。夸张地说，它就是瓦尔泽在巴斯克城散步时发生的那次偶遇，它与周遭的具体环境无关，而与基础设施、建筑、大片城市空间的整体变化相连。

毕尔巴鄂博物馆的例子说明，博物馆“类型学”远比其他因素更能在室内外之间、在主要功能与“表征”内涵之间构筑起一道鸿沟，使

the fact it is "art." The work is isolated from everything that would detract from its own evaluation of itself. This gives the space a presence possessed by other spaces where conventions are preserved through the repetition of a closed system of values. Some of the sanctity of the church, the formality of the courtroom, the mystique of the experimental laboratory joins with chic design to produce a unique chamber of esthetics. So powerful are the perceptual fields of force within this chamber that, once outside it, art can lapse into secular status. Conversely, things become art in a space where powerful ideas about art focus on them.

– Inside the White Cube. The Ideology of the Gallery Space, Brian O'Doherty, *Studio and Cube*, 1999 (essays published from 1976 on the magazine *Artforum*)

In 1976, in the pages of *Artforum* magazine, Irish artist Brian O'Doherty proposes a reflection on the evolution of exhibition spaces, in particular from the advent and proliferation of anodyne white halls as main spatial support in the exposition of artworks. According to the artist, we passed from the simple frame as an edge separator between a work and another, to the need to set the same work within a space as neutral as possible that could create a pneumatic vacuum around it, so the visitors fully concentrate their attention on the work.

This has transformed the presence of architecture, giving to the support walls of the work an abnormal importance, removing it from the artworks, transforming the museum into a space no longer subjected to the artist, but equal to it, if not more important. If this reflection was valid in 1976, especially for galleries and more specifically for the interior of the museum, the functional deformation of the architecture of the museum today has extended to the wrapping, to the external appearance of the building. The two most famous – Guggenheim Museum, New York and Bilbao – well embody the evolution; if the Wright's exhibition path forces the hosted artists to challenge the space enclosed by a continuous ramp, Gehry's museum is absolutely uninterested in its main function, the exhibition. The Spanish museum is designed for other purposes: the building is transformed into a bearing episode of the new city plot, a focal point within the urban growth of the city. To do this, Gehry expands immeasurably the scale of art, considering not a specific place, but the city as a whole. In order to build a rich path – "going, walking" – he needs to turn the museum into a work of art in itself. The Guggenheim Bilbao is an urban accident; it is the meeting, exaggerated, of Walser while walking in the Basque city; it does not have a relationship to its specific surroundings, but it finds a connection with the general transformation of infrastructures, buildings, large urban voids.

The example of Bilbao tells how the museum "typology", far more than others, has been able to afford a total gap between interior and exterior, between the primary function and its connotation of "representation", giving rise to some formal adventures that try to scream their presence throughout the urban surroundings.

Some of the museums presented here strongly follow this trend: the Soumaya Museum is configured as an alien object, unbal-

索玛雅博物馆顶层的多功能画廊
multi-use gallery on the top floor, Soumaya Museum

FREE–Fernando Romero建筑公司提供
courtesy of FREE-Fernando Romero

安达卢西亚历史博物馆庭院中的环形坡道
courtyard circular ramps, Andalucia's Museum of Memory

Javier Callejas建筑公司提供
courtesy of Javier Callejas

一些急切想要在城市环境中展现自我的形式探索创新异军突起。

本文介绍的一些博物馆紧紧抓住了这一趋势：索玛雅博物馆就像一个失去平衡的天外来物，建造它所投入的巨大努力应当引起人们的关注，这种努力可以让博物馆的魅力扩大到几百米之外，让整个城市都感受到它的存在。同样极富魅力的还有广东省博物馆，它就像一块人造巨石，表面上刻有大片的彩色割痕，好似一个身负重伤的巨人，虽然与周围的景物格格不入，但富有磁性，格外迷人。安达卢西亚历史博物馆则凭自己符合欧几里得原理的外壳以近乎神秘的方式吸引公众的视线；它是一个极小的外壳，可以在严格的形状内增强一切可能出现的内部张力，创造出一套几何秩序，而一片崭新的城区就以此为基础落成。

走，慢慢走……看云卷云舒

2006年，格哈德•里希特在一本厚厚的《图集》中收录了他在四十多年间搜集的图像，以一种分类档案的形式展示“发现的艺术品”，这同时也构成了他本人的一部人生历程，其私生活与职业生涯并无分别：一套连续数百张的照片，往往难以区分，但由于数量众多，便能够还原那种连续创作的感觉。

在奥多尔蒂阐释自己批判白立方理由的同一年，里希特用一系列名为“云朵”的照片找到了解决方案，这些照片中呈现的是数十朵浮动的云彩，它们的形状在视野内不断变幻。

此后，这位德国艺术家用素描向人们展示了各种可能的展厅形状，他为这些作品取名为“房间”，人们可以从中看到改良版的白立方。按照奥多尔蒂的说法，新版白立方被一望无际的云朵割裂开来，算是对博物馆外观的一种解放，通过与外界建立友好的关系来打破艺术品四周的窒息氛围。

里希特能够在现代博物馆与世隔绝的世界中寻找到一条出路，假定将某一区域中的游客带到他的微缩世界中，他们就能感受到瓦尔泽所描绘的“走，慢慢走”的意境。

以下是圣特尔莫博物馆的建造过程——在城市道路的基础之上添加一套全新的、复杂的室内路径，在表面通过一大段台阶与位于石山和历史广场这一自然与人为的结合处的新大楼连接在一起。不仅如此，在Côa谷艺术考古博物馆内，仿照里希特的作品，设计者将整座建筑与自然景观完美地融合到了一起，正如那位德国艺术家所创作的“房间”一样，分裂开来，敞开怀抱拥抱外面的世界，让耐心漫步的人们体会到“旅途终了，目睹奇观”的喜悦——摆在他们面前的是一条通往云端的路。

anced, a huge construction effort that requires attention and that can catalyze it within hundreds of meters, making its presence felt within all the city. Equally strong is the presence of the Guangdong Museum, a gigantic artificial stone, engraved on the surface with large colored cuts, a giant injured in a body totally detached from its surrounding, yet magnetic, attractive. In this way, the Euclidean casket of the Andalucia's Museum of Memory attracts, in an almost enigmatic mode; a minimal container that can tighten any possible internal tension within strict forms and that create a geometric order around which a new city is built.

Going, walking… and seeing the clouds

In 2006, Gerhard Richter collects in a thick book, *Atlas*, more than forty years of collecting images, in a sort of raisonné archive of "object trouvé", which constitute a body of reference of the artist, without distinction between personal and professional life: a continuous succession of hundreds of photographs, often indistinguishable one from another but able, thanks to their quantity, to restore the sense of a work in continuous progress.

In the same year in which O'Doherty explains his criticism to the white cube, Richter proposes a way out, with the series of photographs entitled *Clouds*, in which are represented dozens of moving clouds, which fill the field of view with always different configurations.

Through sketches, then, the German artist shows possible exhibition configurations, which he called *Rooms*, in which one can see the white cube, described by O'Doherty, ripped by sweeping views of the clouds themselves, in a sort of liberation of the museum, in a rapprochement with the outside world that upsets the aseptic vacuum around the artwork.

Richter is able to find a way out from the self-referenced world of the modern museum, with a hypothesis that brings the viewer within an area, proportionate to his scale, to the "going and walking" described by Walser.

This is how the San Telmo Museum is built, adding to the urban paths the new complexity of the interior walks, but also on the cover, through a large stair that integrates the new building in the exact junction between the nature of the rocky hill and the artifice of the historic square. Even more, recalling the work of Richter, in the Côa Valley Museum of Art and Archaeology, the building perfectly integrates with the natural landscape, opening up and tearing as in the *Rooms* of the German artist, to embrace the outside world, to give to the patient walker the joy of concluding the journey on an uncommon panorama, toward the clouds.

Diego Terna

迪亚哥•特纳，建筑师，网络杂志《PresS/Tletter》中“想象建筑”版块负责人。曾在意大利米兰和西班牙圣塞瓦斯蒂安学习建筑。在一次青年批评比赛——意大利批评家Luigi Prestinenza Puglisi发起的比赛——中获胜之后，他所撰写的文章出现在若干艺术和建筑杂志（《Compasses》《Arch'it》《Exibart》）上。自2005年起，他一直在Italo Rota公司工作，之前曾供职于博埃里工作室。

Diego Terna, architect, curator of the section "Imagined architectures" in the webzine PresS/Tletter. Studied architecture in Milan, Italy, and San Sebastian, Spain. After he won a young critics competition, called by Italian critic Luigi Prestinenza Puglisi, his writings have appeared in several art and architecture magazines (Compasses, Arch'it, Exibart). He has been working for the office of Italo Rota since 2005, after an experience at Boeri Studio.

MACRO当代艺术博物馆

Odile Decq Benoît Cornette Architectes Urbanistes

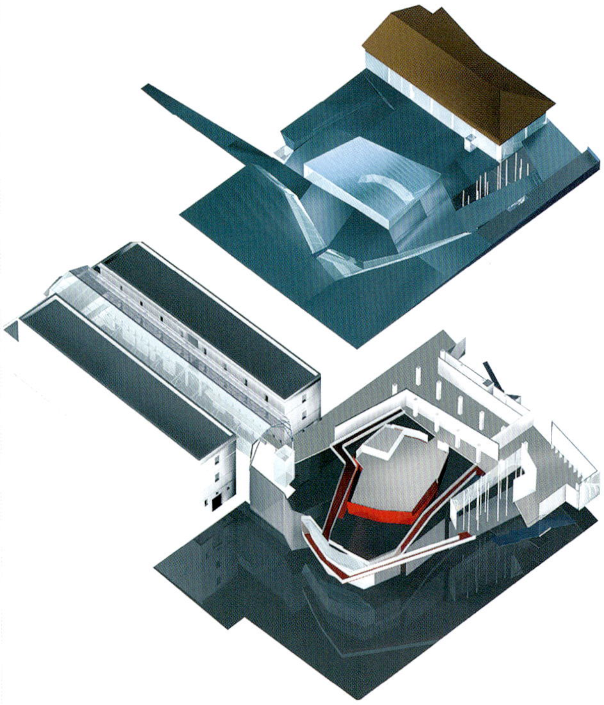

具有美感的领域

将当代艺术画廊引入老式工业建筑中，这个过程具有的复杂性可以使该项目有机会在历史的背景下跨越一体化中的单一姿态。而且罗马的历史背景对于这种类型的介入是一种挑战。

我们的项目融入老式工业产地的复杂性，又面对历史的遗产，具有活力、动态及展示，与该场地现有的静态条件产生了共鸣。

在建筑作品中，剖面图揭示了看不见的部分。它是通过从水平到垂直、从内到外、从门厅到屋顶景观设计的过渡来显现的。新旧展览空间及其相关活动，现代与当代场所都在同一时间被交织在一起，并显现其各自的特性。在这个过渡的体系中，新的干扰物都被布置在周边，使得每一部分都具有复合清晰度，变得有吸引力。

剖面进展过程向三维的转换扰乱了项目每一层的雕塑性外观。屋顶和地板是有刻纹的、断裂的、开放的，构造运动彰显出它们的活力，这种相反方向的位移运动，同时将内外空间联系在一起。

屋顶表面也是一处景观，它映射了下面的艺术活动，创造了一种室内/室外的抽象连续性。这一景观服务于博物馆的游客和这座城市。游客脚下的材料具有多种多样的质感：粗糙的、纵深的、光滑的、柔软的、密集的、明亮的、亚光的。另外，这座景观花园也可供露天展览使用。

会议厅是一个神秘的、形状随意的彩色空间，其特点是设计新颖、密集性高。中央处的活水可以净化空气。天气炎热时，门厅中央的水通过蒸发作用给人带来清新的感觉。此外，空间上方的玻璃墙可以打开，这样门厅里的空气就能排到室外。此外，空气沿着活水系统流动，使门厅的通风透气性变得良好。与周边有活力的场所相比，可预见展览空间及其活动会更加中规中矩，艺术家们将其设计在一个非正规的简单空间内，并提供了多种展览方式：悬挂的、悬垂的、浮动的、在地面上或嵌入地面的。设于展览会内部的坡道和通道为欣赏艺术作品提供了多种可能性。

MACRO Museum of Contemporary Art in Rome

Sensuous territories

With the complexity given by the introduction of a Contemporary Art Gallery in an ancient mixed industrial building, the project is an opportunity to transgress an univalent attitude of an integration into a historical context. Moreover, the Roman historical context is a challenge for this type of intervention.

Inserted into the complexity of an ancient industrial estate and confronted with a historical patrimony, our project is resonating to the existing static condition of the site with dynamism, movement and evidence.

In the architectural work the section reveals the non-visible. The section is shown through the translation from horizontal to vertical, from inside to outside, from the foyer to the roof-landscape-garden. The old and the new, the exhibition spaces and the other activities connected with it, the places for the modern and places for the contemporary are articulated and specifically distinct at

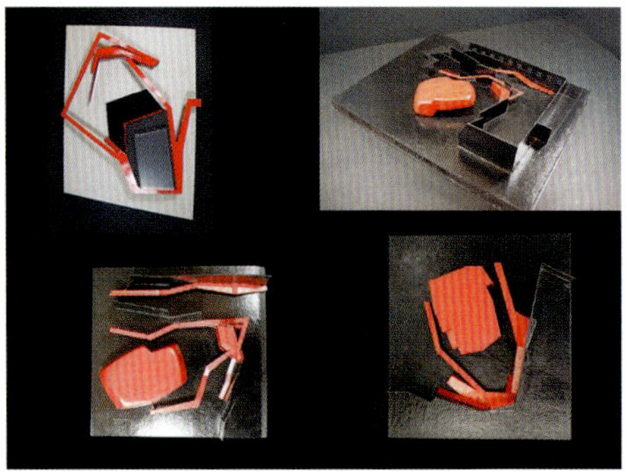

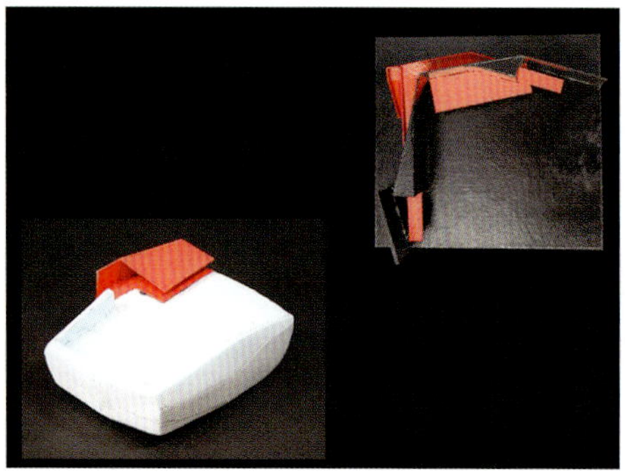

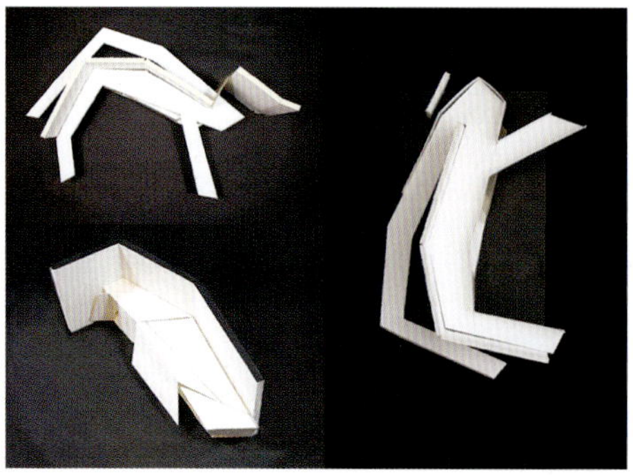

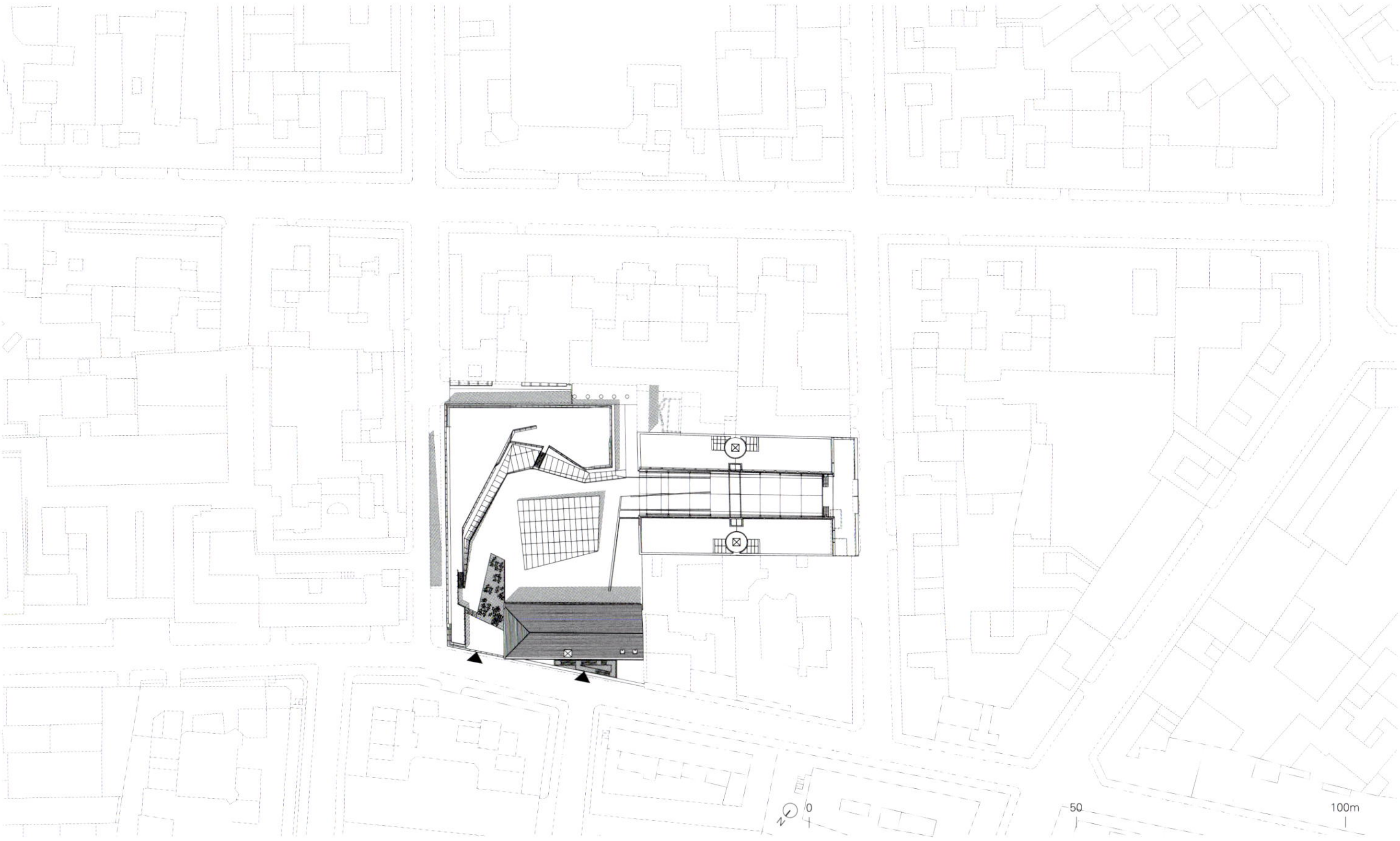
N
0
50
100m

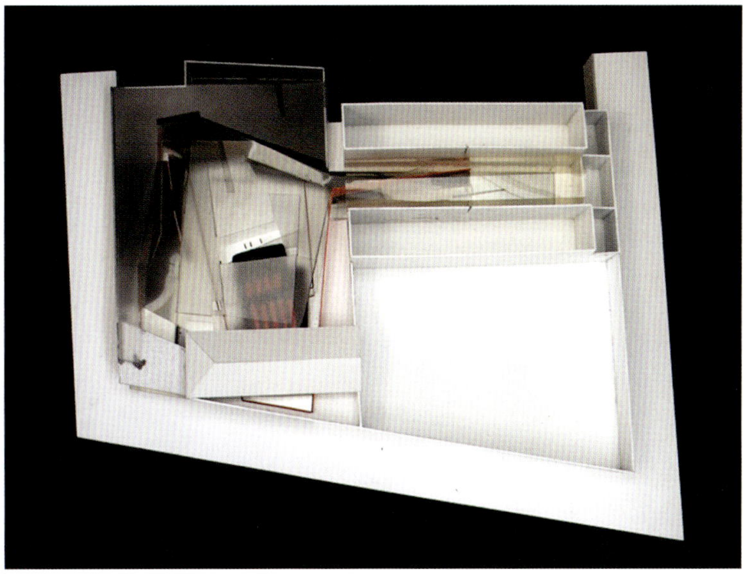

BIRRA
PERONI
GHIACCIO

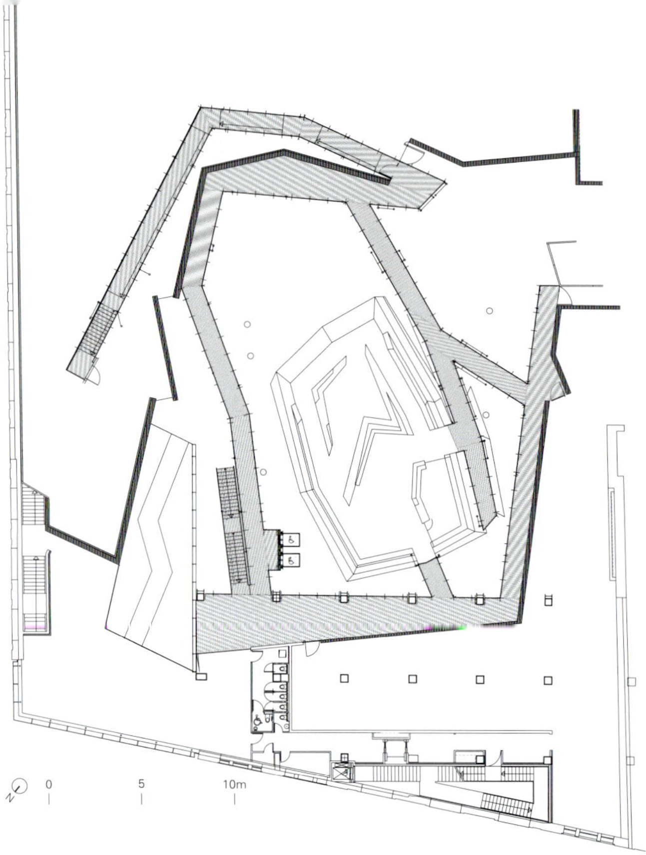

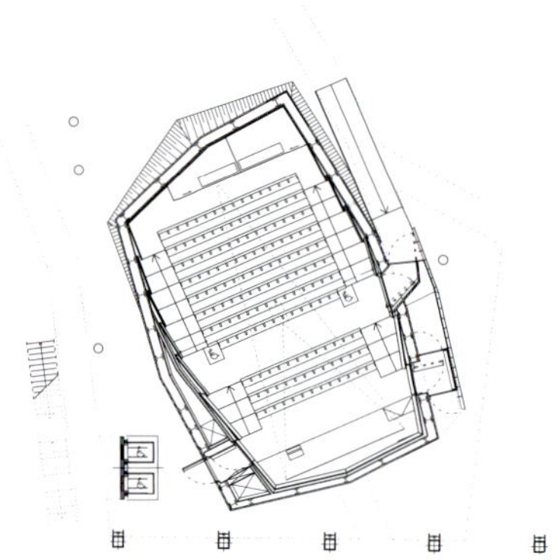

教育大厅地面
educational hall floor

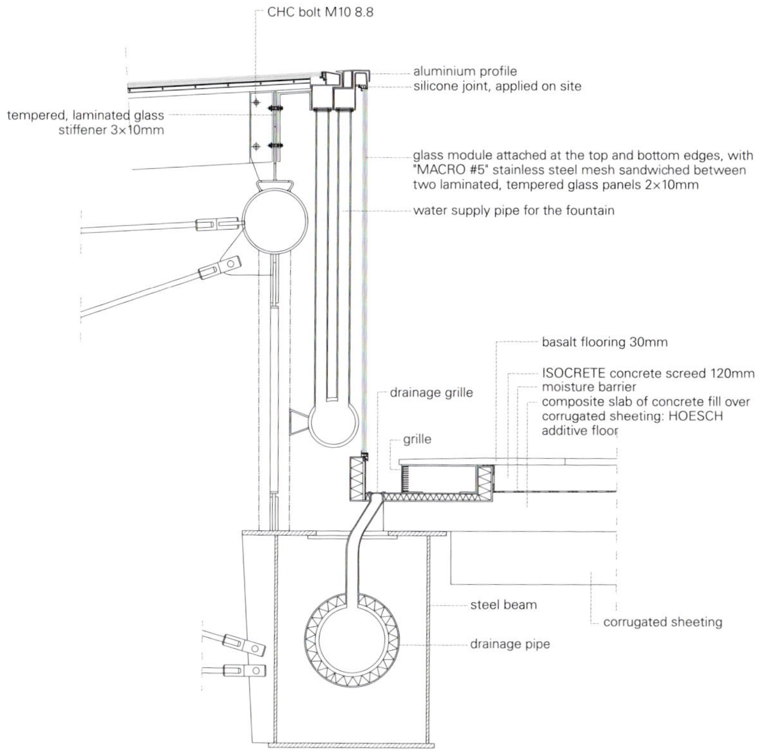

详图1
detail 1

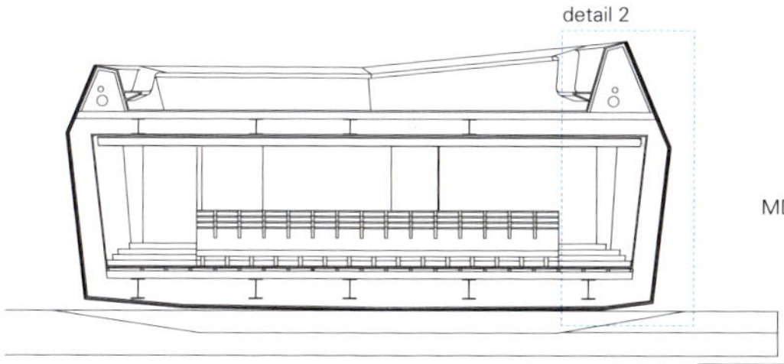

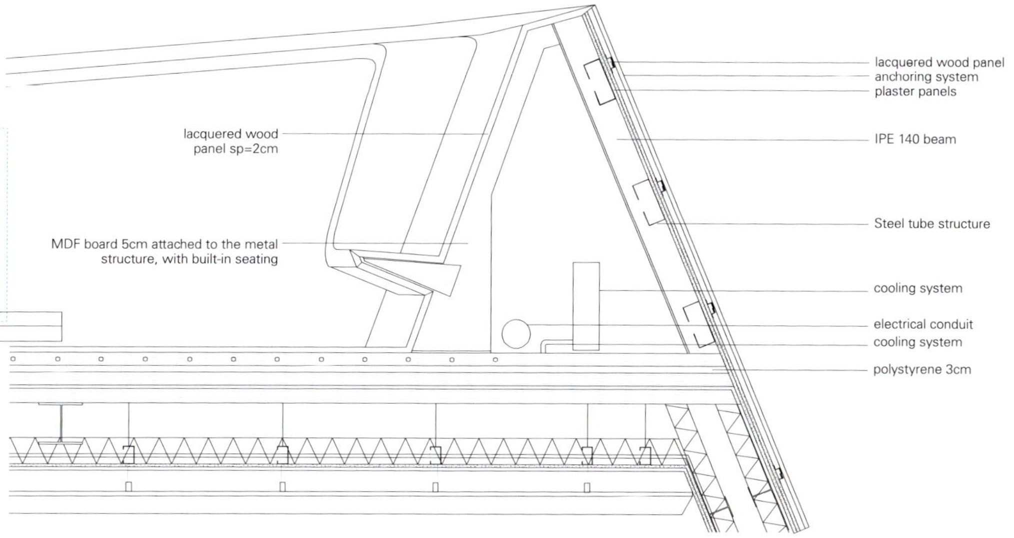

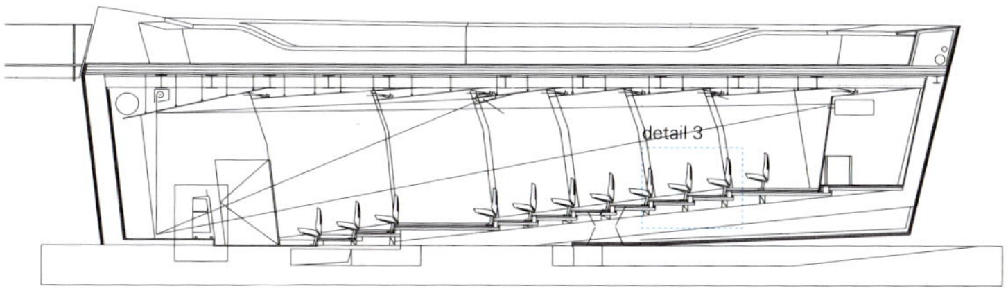

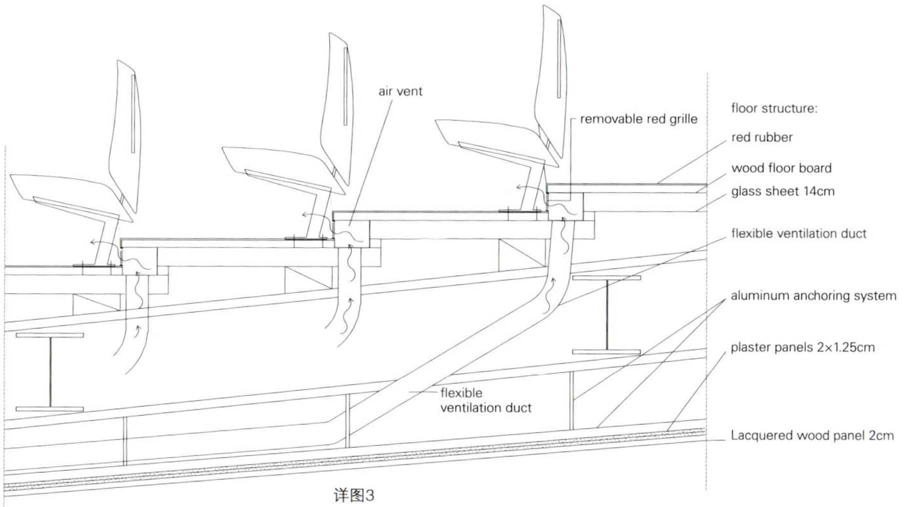

详图3
detail 3

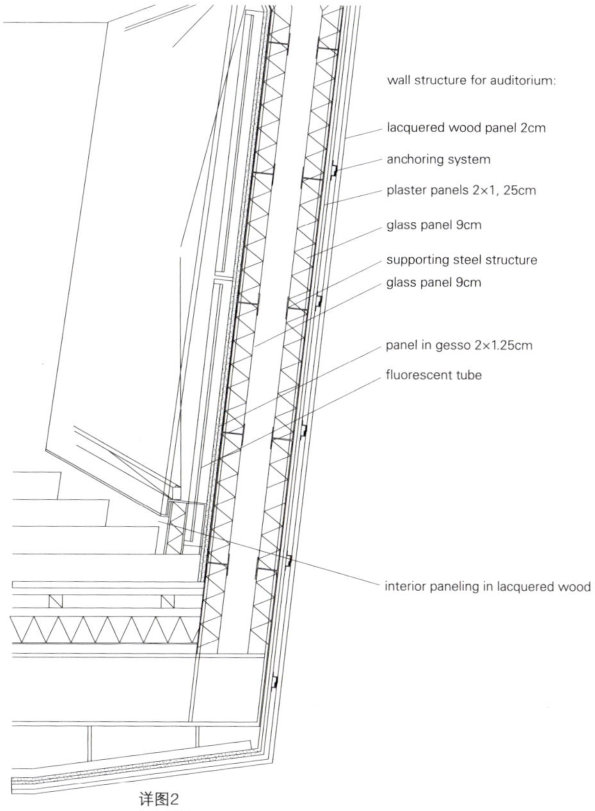

详图2
detail 2

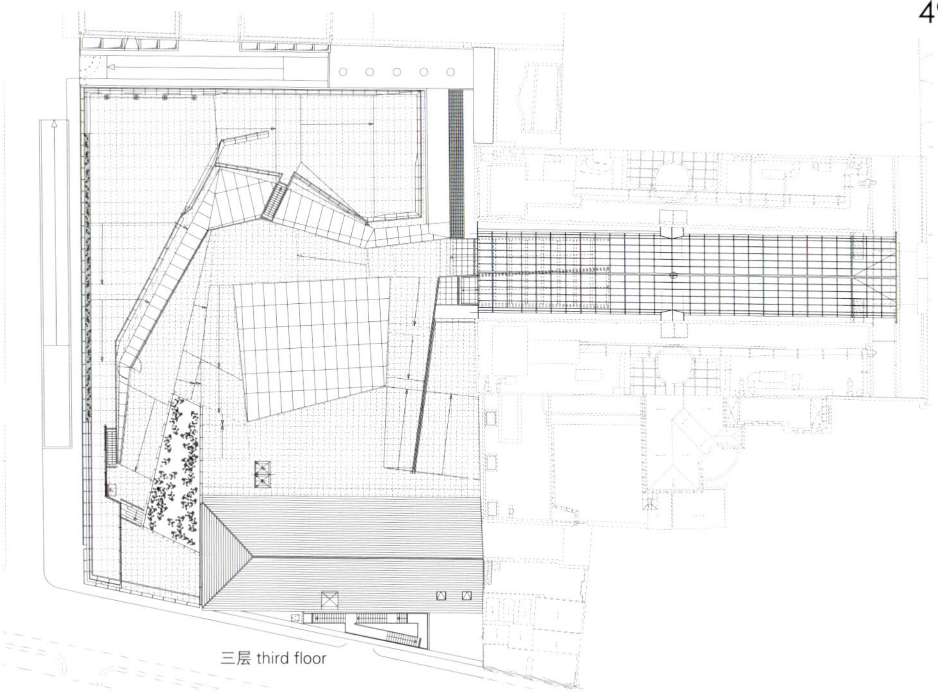

三层 third floor

项目名称：MACRO Museum of Contemporary Art in Rome
建筑师：Odile Decq Benoît Cornette Architectes Urbanistes
项目经理：Giuseppe Savarese
照明工程师：Luce plan
门把手制造商：Valli&Valli
承包商：CDG
项目地点：Rome, Italy
基地面积：12,000m²
竣工时间：2010. 5
造价：EUR 22,000,000
摄影师：courtesy of the architect (©L. Fileticì) - p.47 top
Courtesy of the architect (©Georges Fessy) - p.48, p.49, p.51
Courtesy of the architect (©Poltrona Frau) - p.47 bottom
©Roland Halbe - p.42~43, p.44~45, p.46, p.53

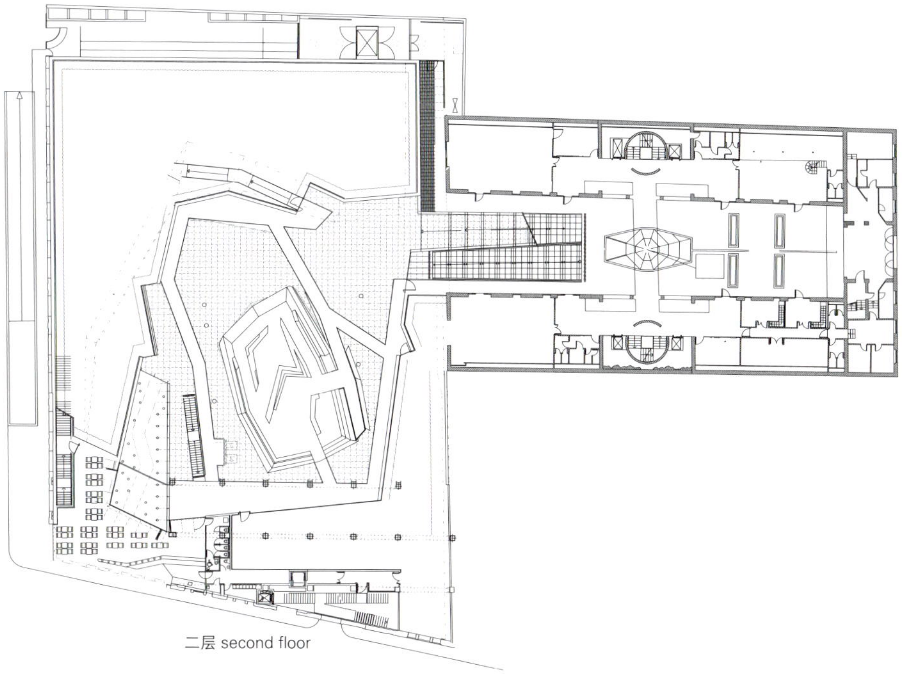

二层 second floor

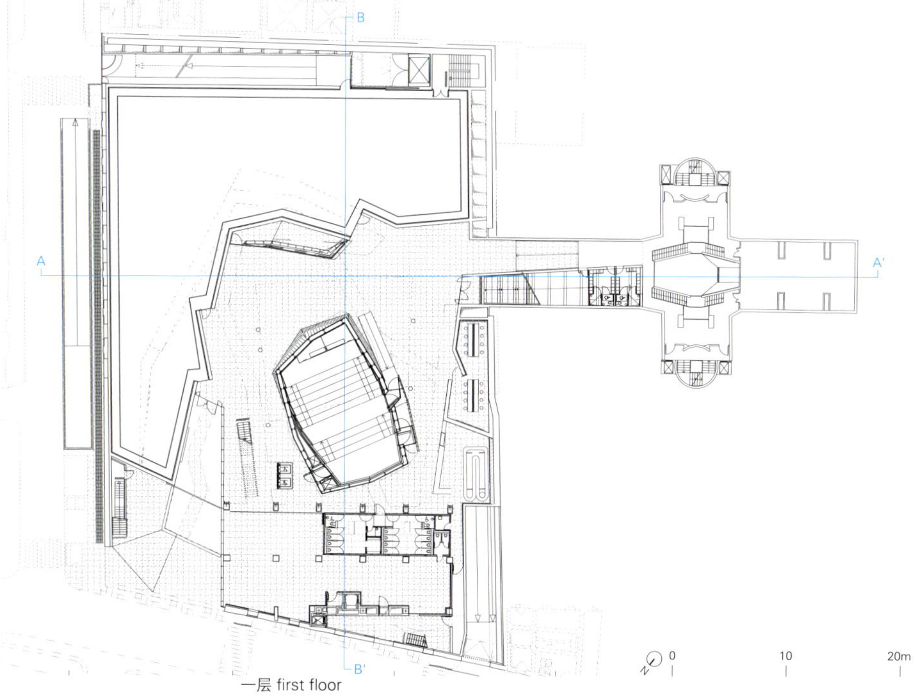

一层 first floor

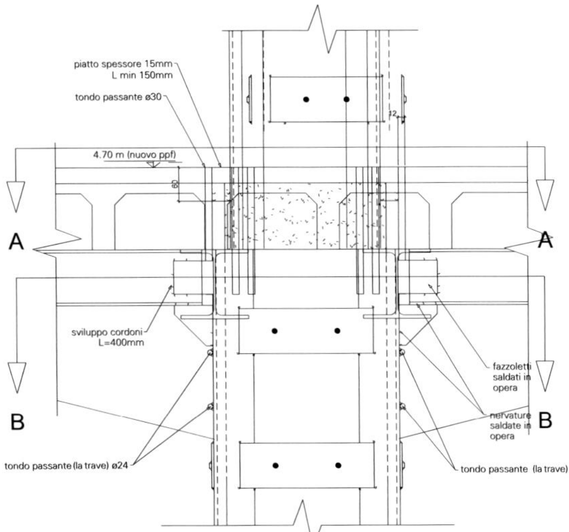

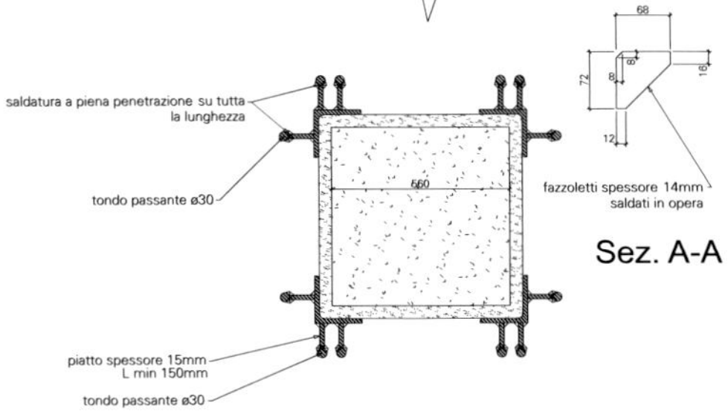

Sez. A-A

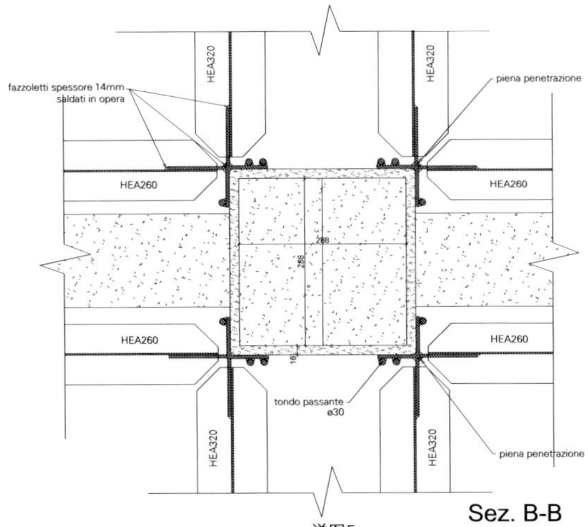

Sez. B-B

详图5
detail 5

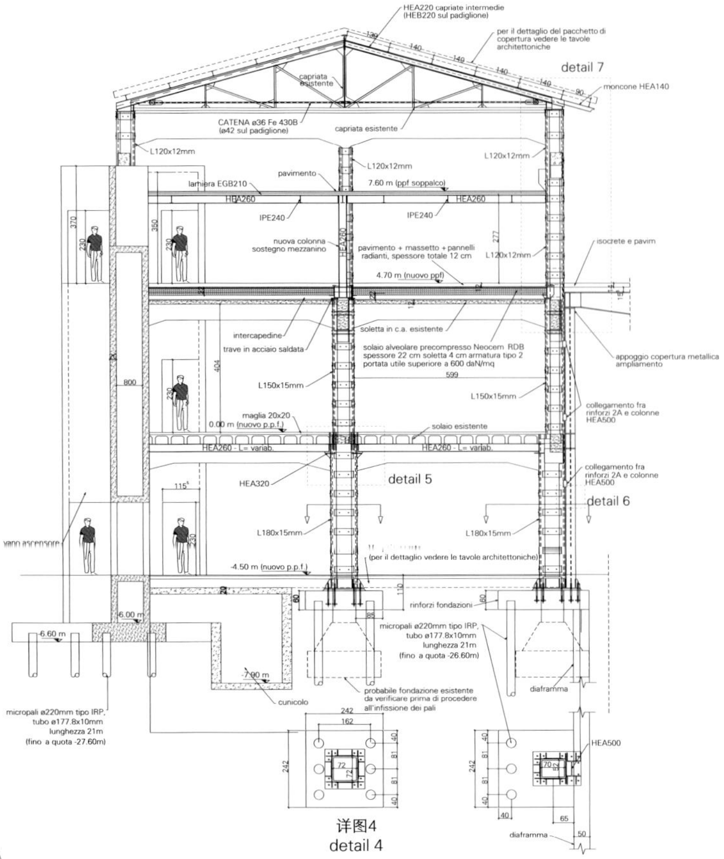

详图4
detail 4

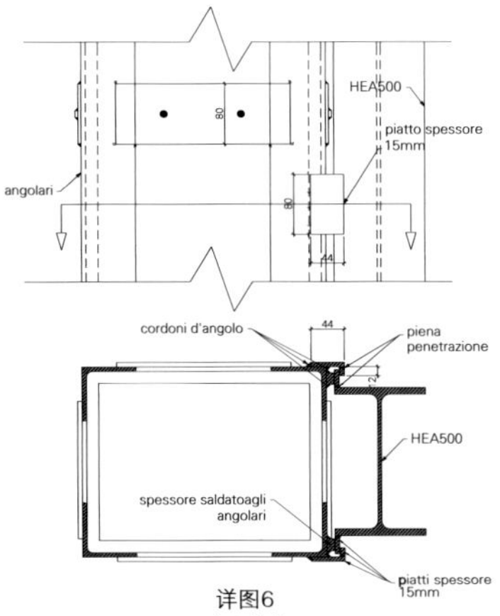

详图6
detail 6

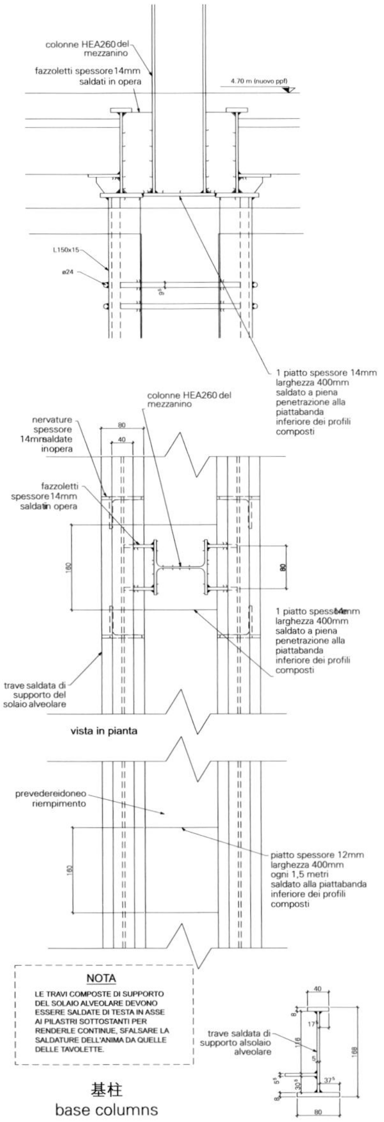

基柱
base columns

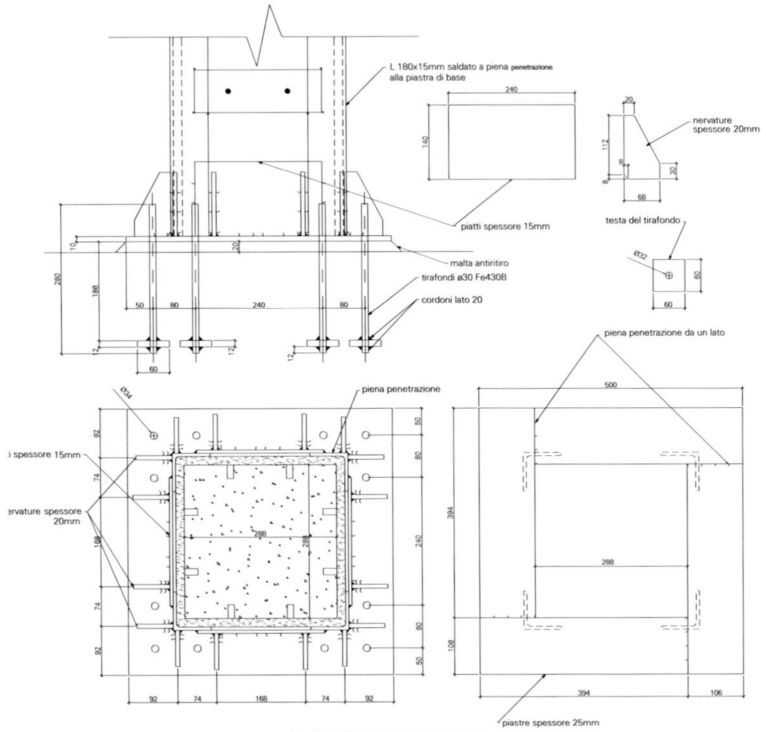

特殊设计的中间层基柱线图
special design of the plinth drawn intermediate

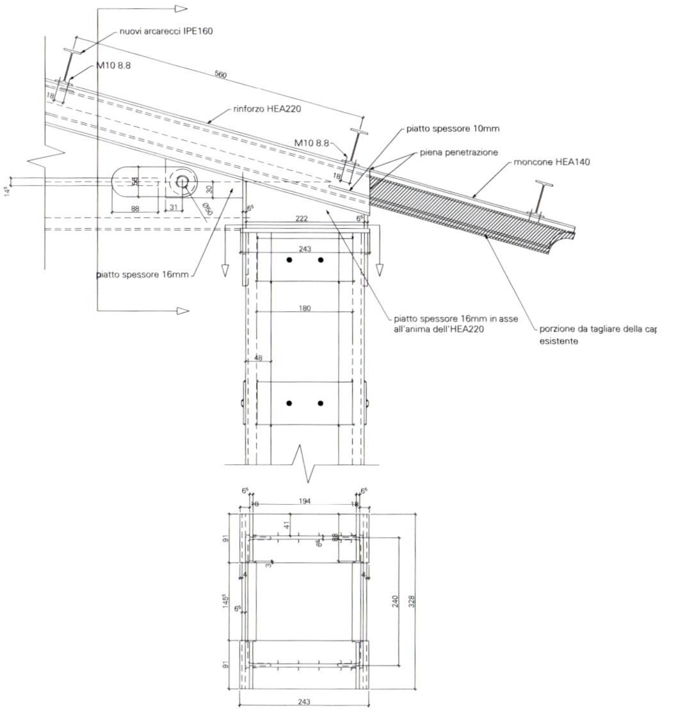

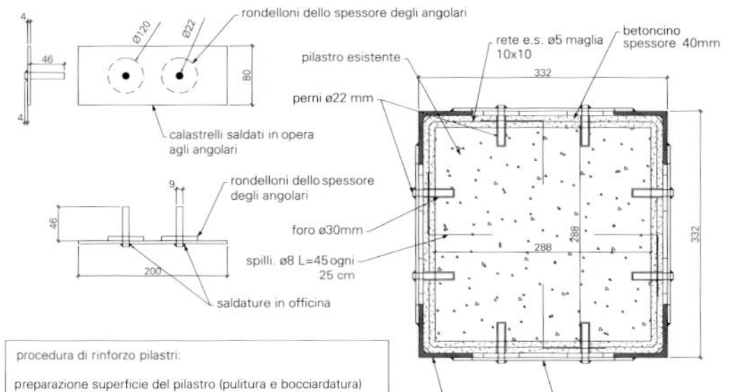

procedura di rinforzo pilastri:

preparazione superficie del pilastro (pulitura e bocciardatura)
inghisaggio spilli con resine
applicazione rete elettrosaldata
applicazione betoncino
foratura con carotatrice a tazza, foro ø30mm profondità 120mm
applicazione profili a L e loro solidarizzazione alle fondazioni
inghisaggio perni ø22mm (preventivamente saldati ai calastrelli)
saldatura in opera dei calastrelli agli angolari
finitura delle superfici

钢筋柱（典型）
reinforcing pillars (typical)

angolari piano terra: L180x15
angolari primo piano: L150x15
angolari secondo piano: L120x12

nota: gli ingombri fuori tutto sono riferiti ai pilastri del piano interrato stilata intermedia.

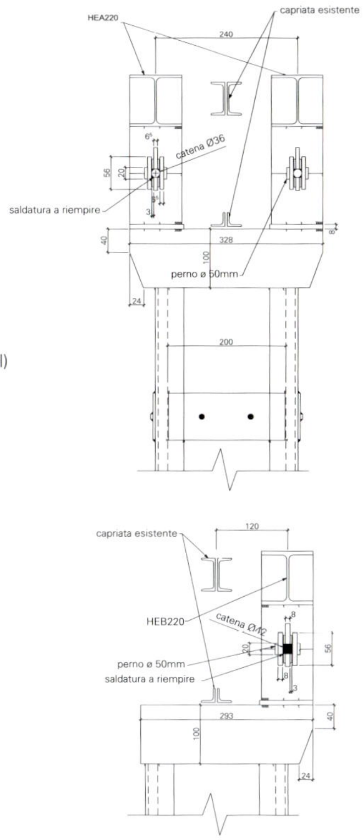

连接新的钢梁——加固的柱
connecting new steel beams
-reinforced pillars

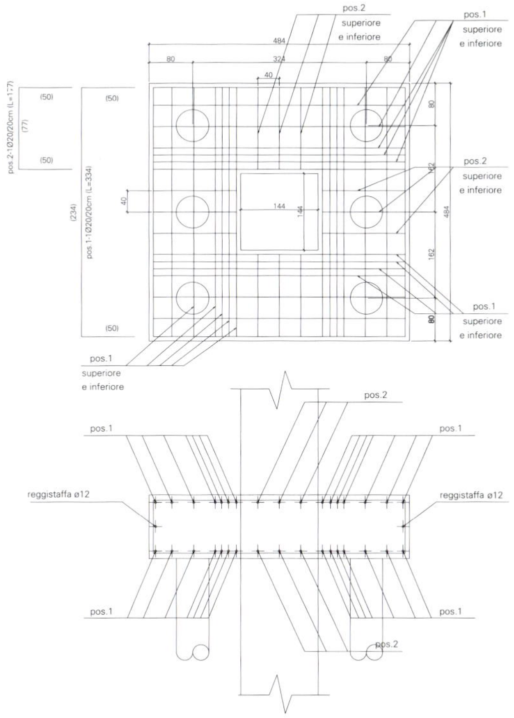

钢筋柱（典型）
reinforcing pillars (typical)

ELEMENTO GRECATO EGB210 - spessore 8/10 (o equivalente)
RETE ELETTROSALDATA ø6 MAGLIA 15x15 cm sovrapp 2 maglie
CALCESTRUZZO Rck=350

SOVRACCARICO UTILE >500 kg/mq
PERMANENTI PORTATI VARI: 10 kg/mq
PESO PROPRIO CLS: 190 kg/mq
PESO PROPRIO LAMIERA: 6.28 kg/mq
INCIDENZA CLS: 0.072 mc/mq

特殊的共享板面
particular share deck

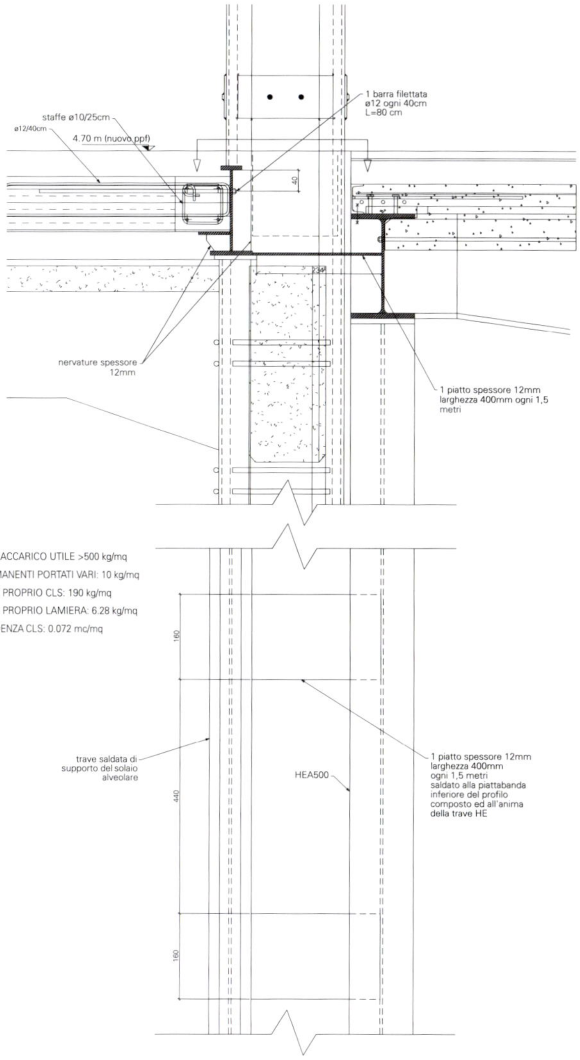

详图7
detail 7

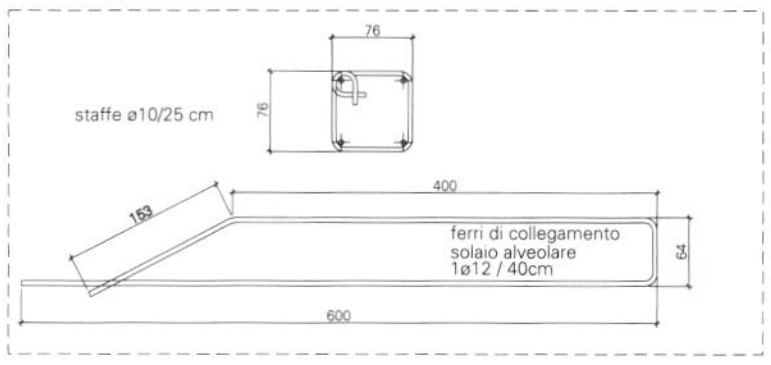

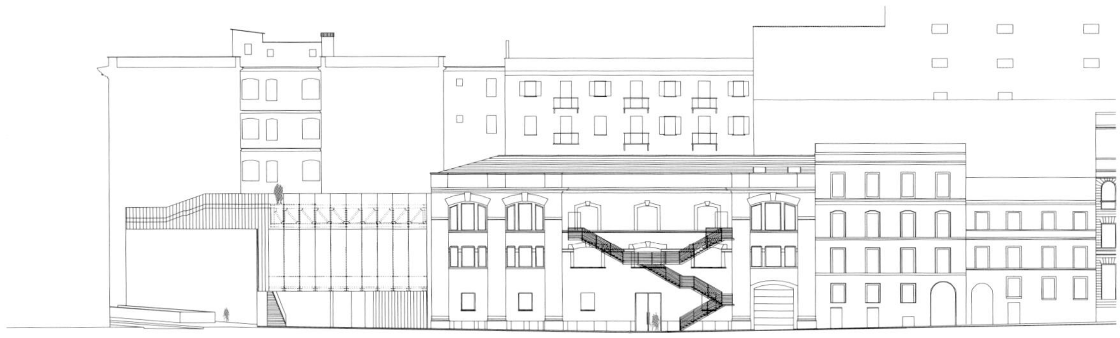

西北立面 north-west elevation

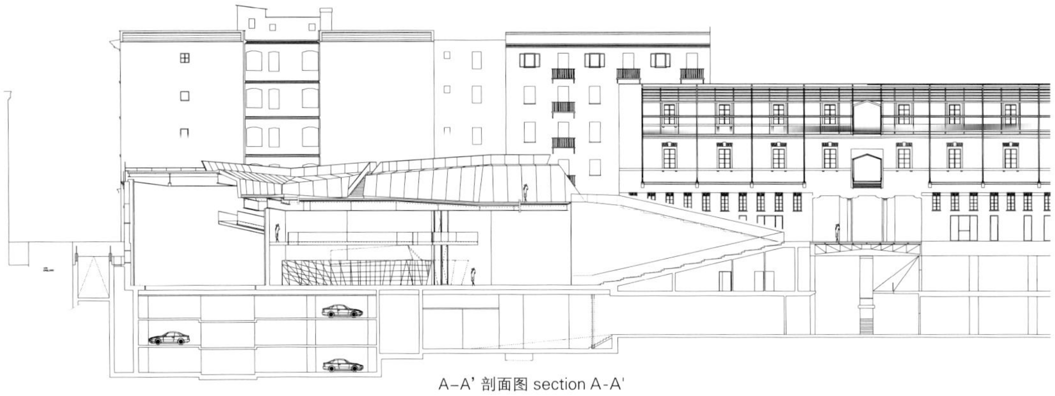

A-A' 剖面图 section A-A'

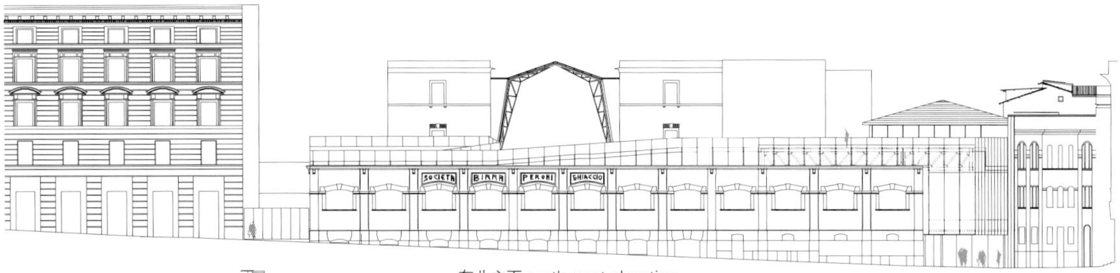

东北立面 north-east elevation

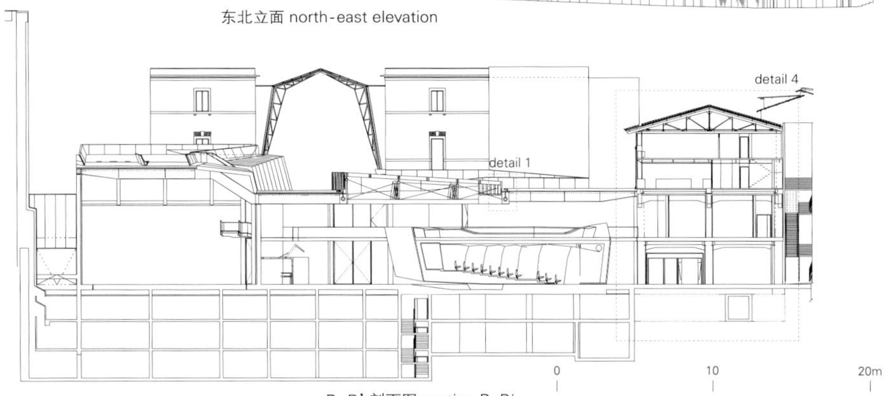

B-B' 剖面图 section B-B'

the same time. In the system of transition in which the new contaminates are all around, every part wins a complex clarity and becomes attractive.

The transposition in three dimensions of the section process perturbs all the inscription surfaces of the project at each level. The floor and the roof are engraved, fractured, open and animated by tectonic movements which give opposite directions for displacements and, at the same time, connect and articulate inside and outside spaces.

The roof surface is a landscape. It is a projection surface of the art trail below. It creates an abstract continuity outside/inside. This landscape is offered to the visitors of the Museum and to the city. It offers multiple textures under the feet of the visitors: roughness, deepness, smoothness, softness, density, brightness, mattness.

In addition, the landscape garden is a place for open air exhibitions.

The enigmatic free-shaped colored volume is a conference hall. It is identified as unique and with a heavy density. The central volume of running water exists to refresh the air. Over the center of the foyer it creates a sensation of freshness by evaporation during the hot weather. Moreover, the glass sides of the high parts of the volume could be opened. So, the air of the foyer escapes and, by passing along the running water, a light ventilation is created.

In contrast with the dynamism of the spaces around, the exhibition spaces and events could be perceived as more neutral. They are developed in a non-regular, but simple, space given to the artists, offered to multiple ways to exhibit: hanging, suspending, floating, on/in the floor. The ramps and gateway which develop walk inside exhibitions offer different possibilities to look at arts.

Odile Decq Benoît Cornette Architectes Urbanistes

SALE
ESPOSITIVE

圣特尔莫博物馆的扩建

Nieto Sobejano Arquitectos

圣特尔莫博物馆经过了多年持续的修整，其部分的物质与功能特点已经发生了改变，从而形成了其现状。建筑位于城市结构与Monte Urgull地势的交界处，它的位置从另一方面反映了圣塞瓦斯蒂安很有代表性的城市问题：从未得到能够彻底地区分自然和人工景观之间的方案。

现在的问题是如何处理圣特尔莫的扩建，以应对迫切的空间与环境绿化所提出的新要求，并能随着时间的推移，展示它与地理位置的关系。

这种直接与激进的姿态又极其矛盾地暗示着它融入了Monte Urgull景观中。我们将专注于建造一道新型的、具有一定进深的轻质“绿墙”，它由现存的地形定义，并藏身于内部用来安置新方案的两个展馆中。这一决定增强了历史建筑与博物馆新建入口的鉴赏价值，新入口提供了访问旧建筑——包含永久展品——和展示临时展品的新展馆的通道。因此主前厅自然地衔接了几个新的区域，如行李寄存处、商店、礼堂、多媒体资料中心、报告厅和餐厅，它们用这些特质完善了博物馆不可或缺的领域。

一道“绿墙”，从某种意义上暗指了一种建筑理念，涉及到项目的方方面面。因此，稍微改变一下墙的方向就足以使人行道自然地通往Monte Urgull，形成一个露天展览场地或一间向景观与城镇开放的咖啡厅。

新建筑/屏幕的轮廓由围绕整幢建筑的多孔金属表皮显现出来，苔藓、地衣和其他植物包裹着金属表皮，这些植物最终会覆盖整座建筑。这样的设计与天然/人工的联系的表现不尽相同。与艺术家莱奥波尔多•弗兰和阿古丝蒂娜•奥特罗的合作始于一个铸铝件的组合游戏，进而产生了这种设想，这将是公共领域中一次不同寻常的介入，象征着造型艺术与建筑之间共同的行为领域。

圣特尔莫博物馆的新扩建部分将随着季节的变换而更改它的外观：它时而会褪色，与山上的植被混合在一起，时而会重新出现，形成一个长期不加润饰的墙壁；或许意外地映射了建筑与流逝的时间形成的难解关系。

San Telmo Museum Extension

The Museum of San Telmo, in its present condition, represents the result of a long process of successive modifications which has partially altered its physical and functional character over the years. Its location on the fringe where the urban structure meets the topography of Monte Urgull is a reflection, on the other hand, of an urban problem very characteristic of San Sebastián: the solution of a division is never completely solved between natural and artificial landscape.

How to approach a contemporary extension of San Telmo in response to new requirements for space and stringent landscaping conditions, while expressing its connection to the location with the passing of time?

The direct and radical gesture implies paradoxically its practical

项目名称：San Telmo Museum Extension
建筑师：Fuensanta Nieto, Enrique Sobejano
项目设计建筑师：Miguel Ubarrechena
项目团队：Stephen Belton, Juan Carlos Redondo, Pedro Guedes, Joachim Kraft, Alexandra Sobral
结构工程师：NB 35 S.L.
机械工程师：R. Úrculo Ingenieros Consultores, S.A.
建筑公司：U.T.E. San Telmo. Amenabar/Moyua
工地监管：Fuensanta Nieto, Enrique Sobejano, Miguel Mesas Izquierdo (technical architect)
模型：Nieto Sobejano, Juan de Dios Hernández–Jesús Rey
甲方：San Sebastián City Council
地点：Plaza de Zuloaga, San Sebastián
设计时间：2006
施工时间：2007~2011
摄影师：©Fernando Alda

N 0
20
50m

STM
San Telmo Museoa

dissolution in the landscape of Monte Urgull. We will limit ourselves to build a new "green wall", deep and light, which is defined by the existing topography, and which hides in its interior two pavilions which will house the new program. This decision heightens the appreciation both of the historical buildings as well as the new entrance to the museum, which offers access to the old building – which will incorporate the permanent exhibitions – as well as to the new pavilion for temporary exhibitions. The main vestibule will therefore constitute a natural link with the new areas for cloakrooms, shop, auditorium, mediatheque, lecture hall and cafeteria which complete the necessary areas in a museum with these characteristics.

A "green wall": on certain occasions the metaphor associated with an architectural idea gives a sense to each and every aspect of the project. Hence the slight changes of direction of the wall are sufficient to provide a natural solution to pedestrian access to Monte Urgull, to configure an open air exhibition space, or to house a café-terrace open to the landscape and to the town.

Rather an expression of the natural/artificial relation which runs throughout our proposal, the new building/screen will be defined by a perforated metal skin enveloped in moss, lichen and other plant species which finally will come to surround the whole building. In collaboration with the artists Leopoldo Ferrán and Agustina Otero starting with a combinatorial game of cast-aluminum pieces expressively conceived for this occasion, this will be an unusual intervention in a public area which represents a common field of action between plastic arts and architecture.

The new extension of the San Telmo Museum will modify its appearance with the passing of the seasons: it will fade on occasions and blend with the vegetation on the hill, and will reappear on other occasions evoking a long unfinished wall: an unexpected metaphor – perhaps – of the difficult relation which architecture establishes with the pass of time. Nieto Sobejano Arquitectos

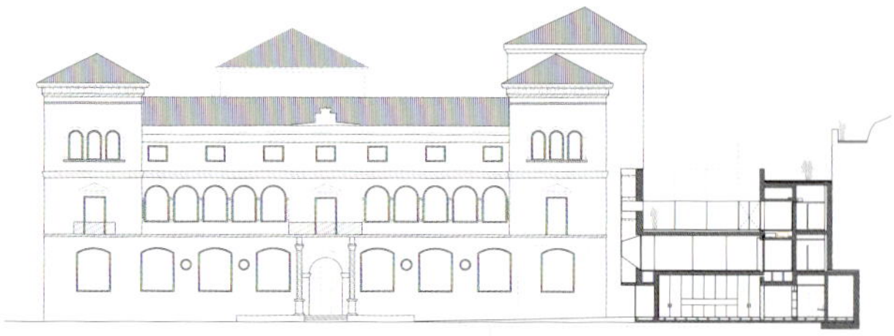

A–A' 剖面图
section A-A'

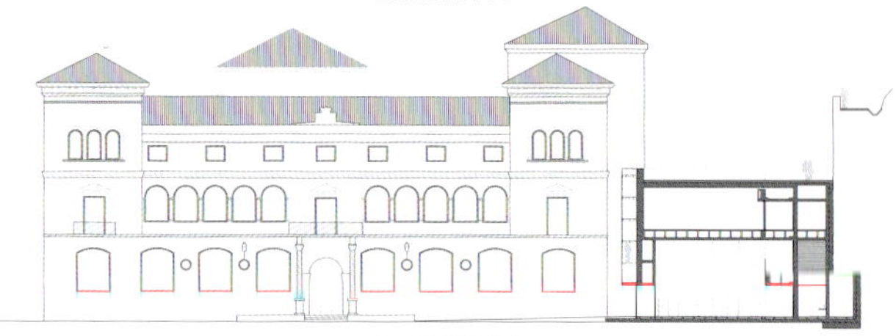

B–B' 剖面图
section B-B'

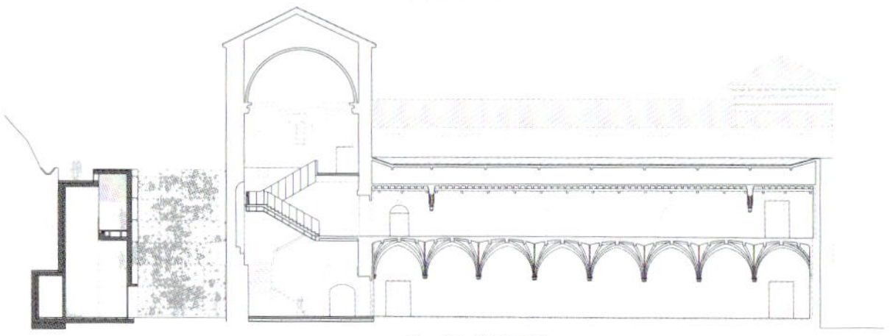

C–C' 剖面图
section C-C'

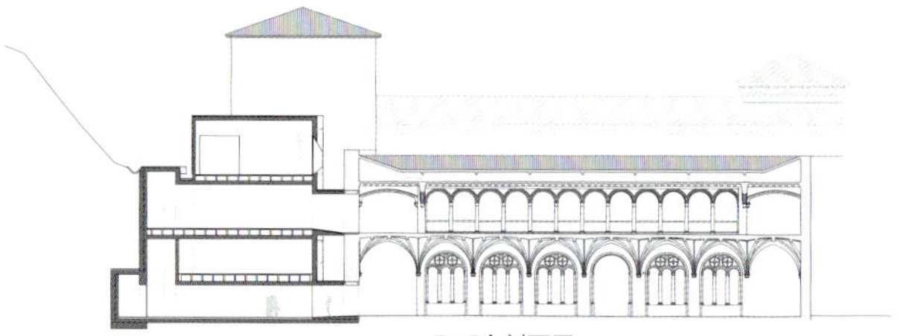

D–D' 剖面图
section D-D'

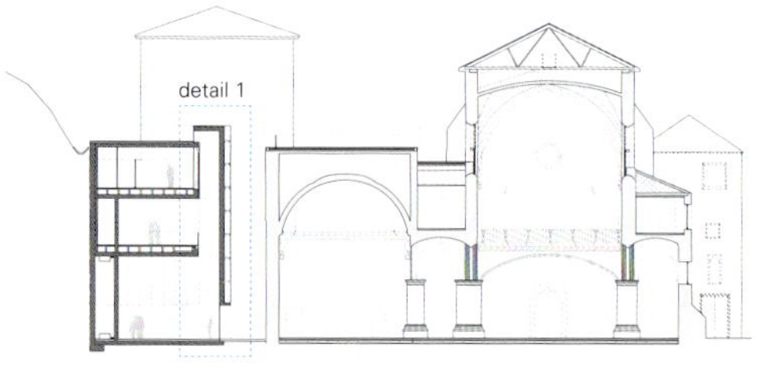

E–E' 剖面图
section E-E'

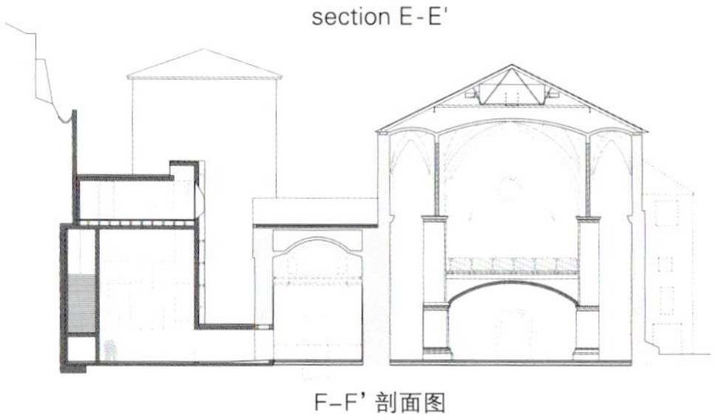

F–F' 剖面图
section F-F'

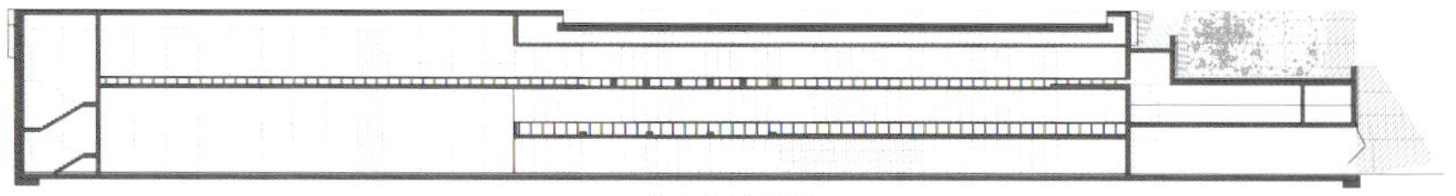

G–G' 剖面图
section G-G'

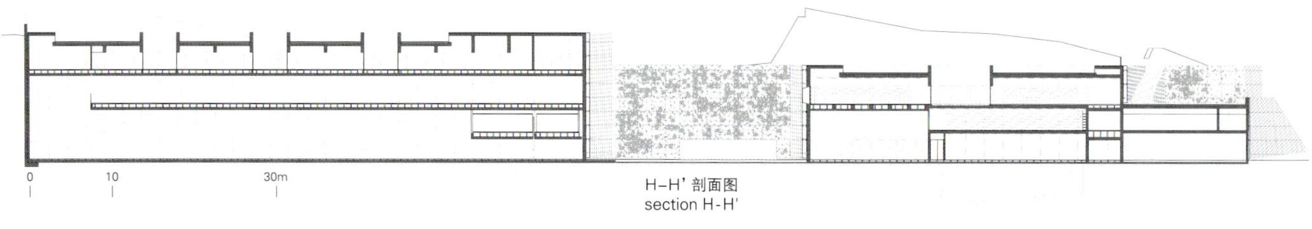

H–H' 剖面图
section H-H'

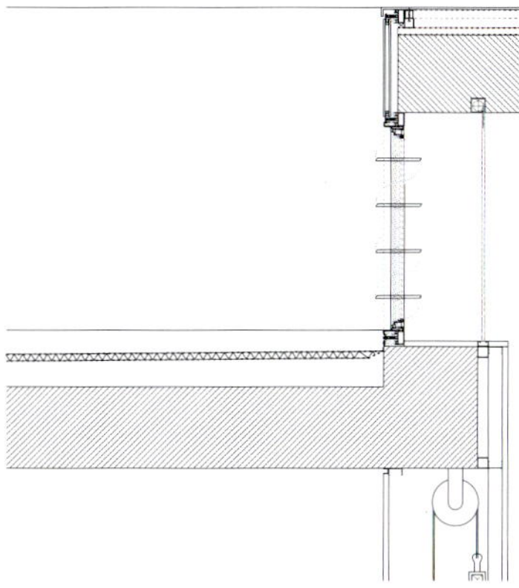

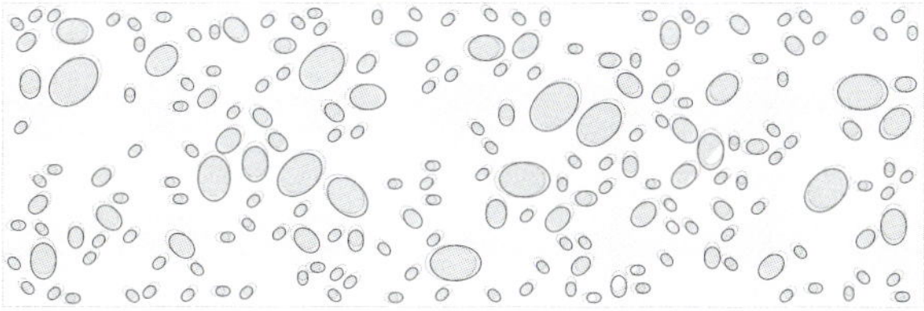

镶板1
panel 1

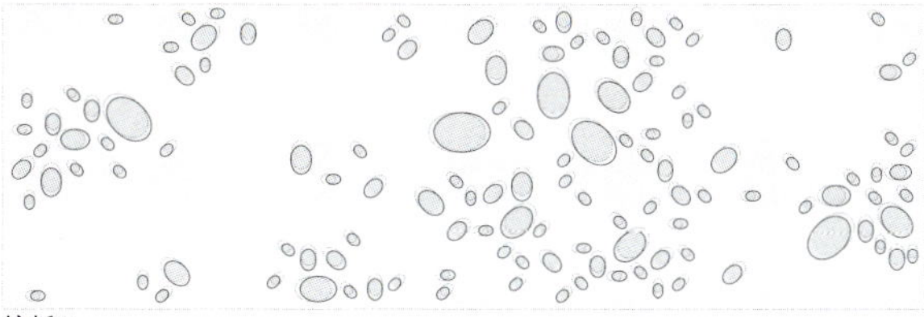

镶板2
panel 2

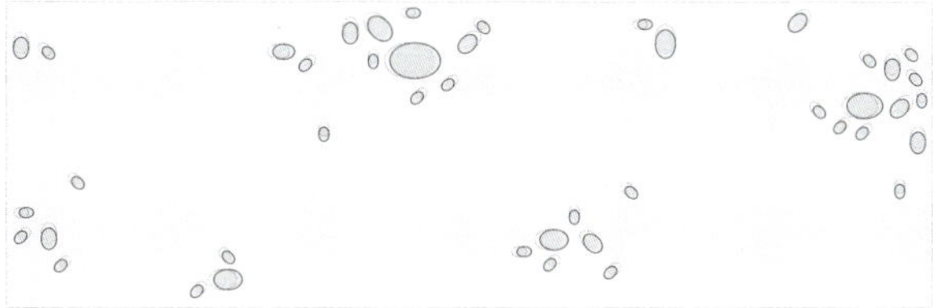

镶板3
panel 3

镶板4
panel 4

镶板楼板
panel floor

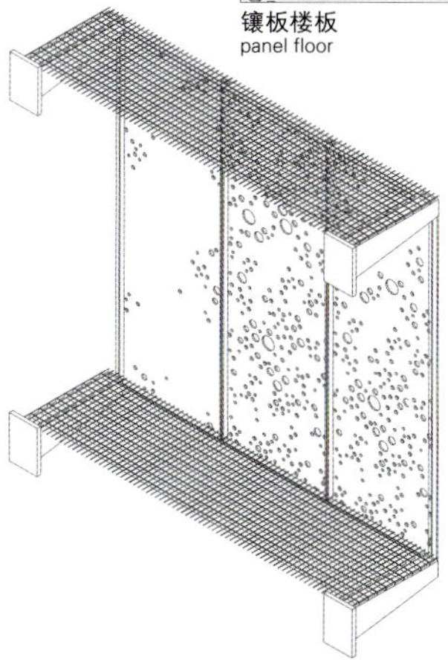

CANTIDAD DE AGUJEROS POR PANEL

ORIENTACIÓN	PANEL 1	PANEL 2	PANEL 3	PANEL 4
	80	36	13	3
	52	42	11	4
	40	21	9	2
	29	28	11	3

DIMENSIONES DE AGUJEROS (mm)

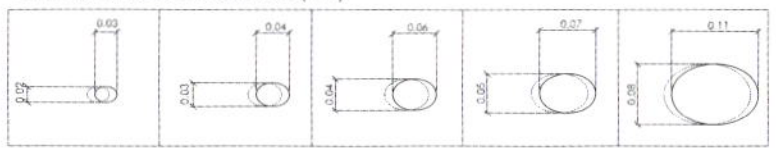

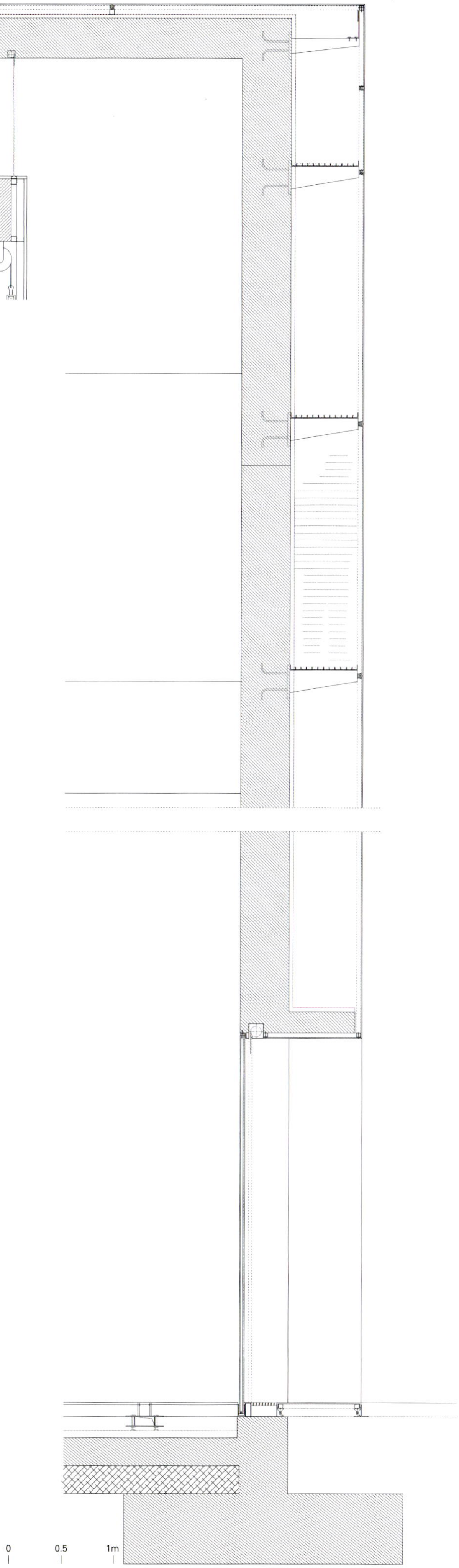

详图1
detail 1

ВОСПОМИНАНИЯ
ecuerdos
de infancia
ko ametsak
childhood
dreams

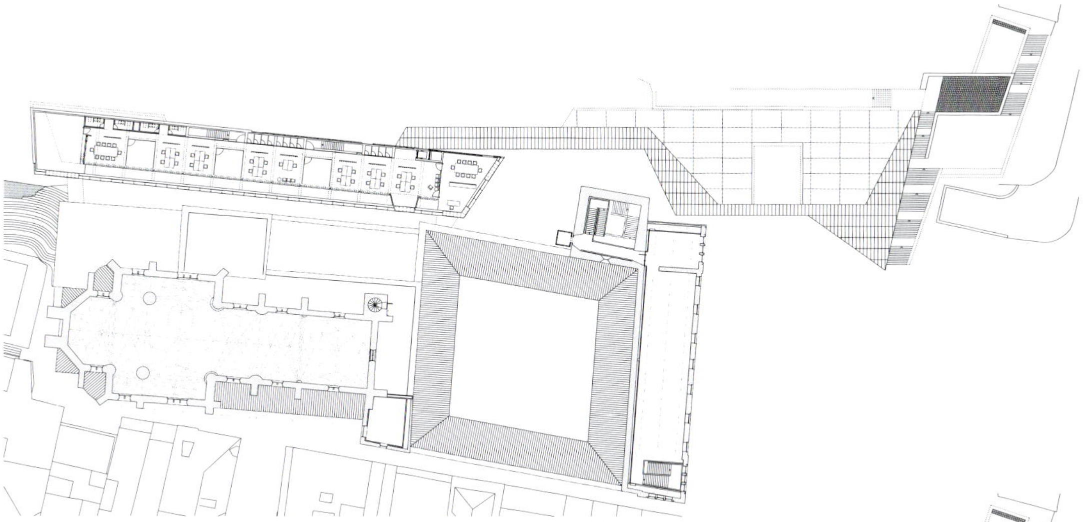

四层
fourth floor

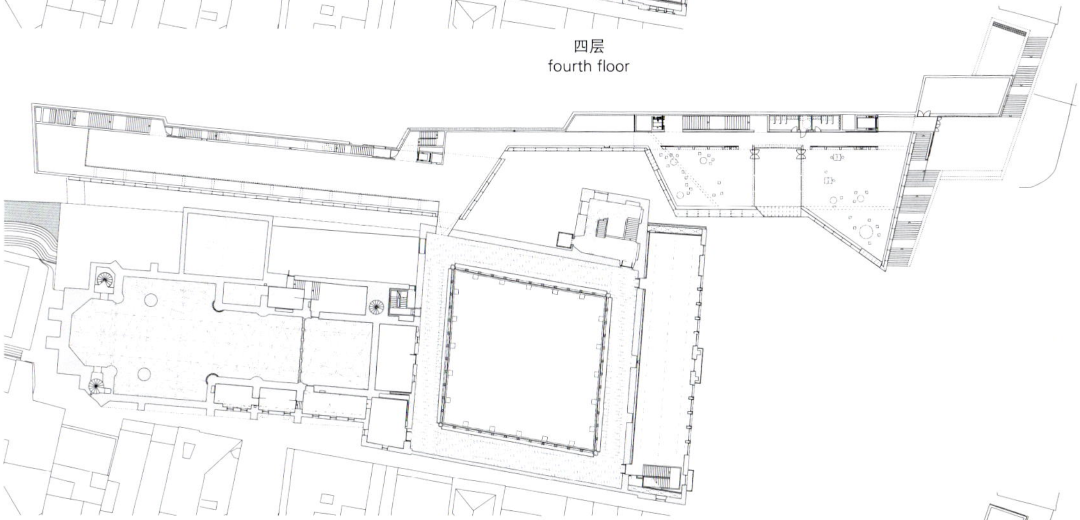

三层
third floor

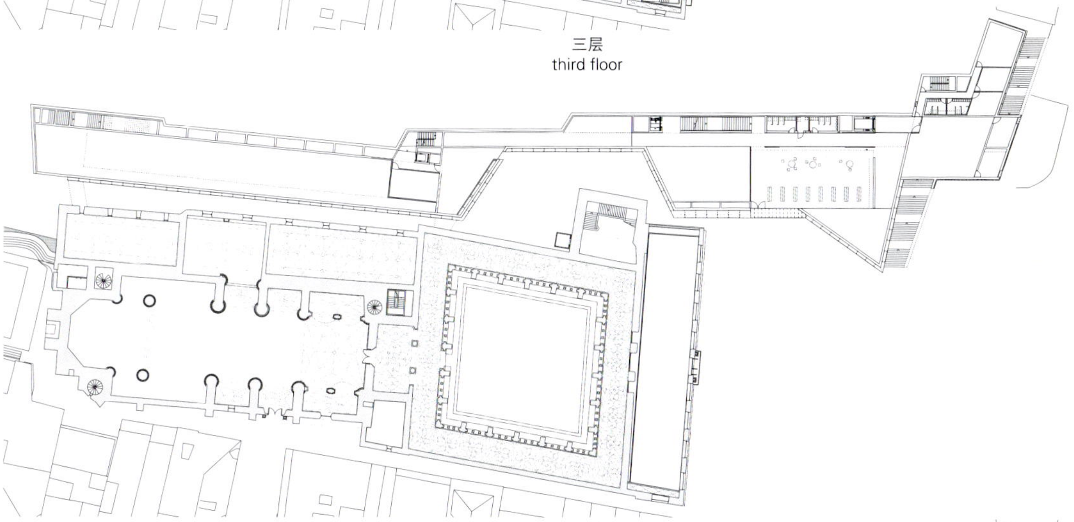

二层
second floor

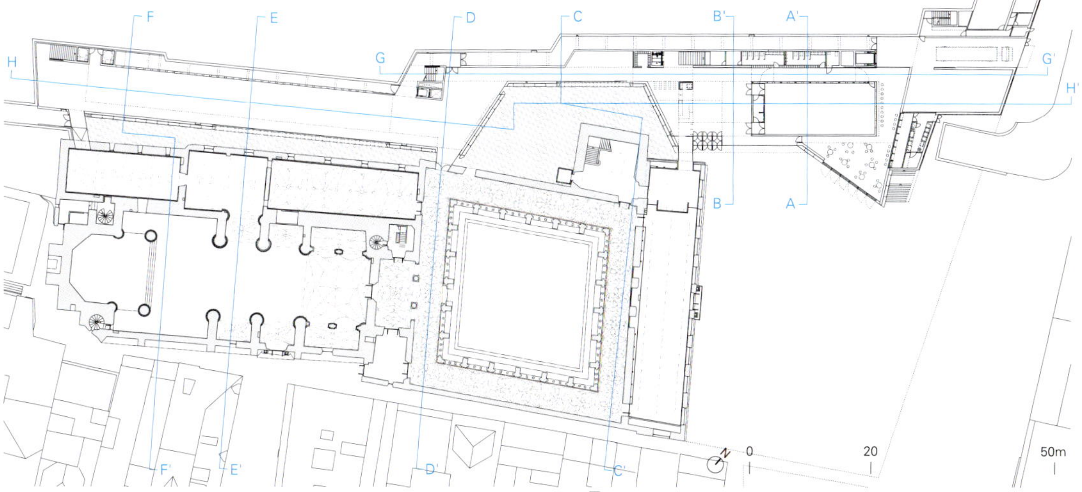

一层
first floor

Côa谷艺术考古博物馆

Camilo Rebelo + Tiago Pimentel

为了在Côa谷修建艺术考古博物馆，我们结合了不同的因素：地形、可达性以及布局规划。以上这几个方面因素的结合对使博物馆融入景观这一概念定义起到至关重要的作用。

地势是一个决定性的因素，因为我们必须要应对博物馆入口和建筑物内部之间巨大的差异性。抵达处选择在了该地势的顶部，成为创建一个平台，一个台阶，一座能看到山脉和峡谷庞大的、令人印象深刻的景观的观景台的原因之一。抵达处很安静，是一个停车的地方，同时也是博物馆的入口。就地形而言，该结构策略是因地制宜的；这意味着该平台与延伸到建筑的道路处在相同的水平线上，同时地形沿着墙壁的走势向下延伸，在正立面上显示出建筑的总高度。

在地形的较高处，山谷的交汇直接形成了一个庞大的三角形。在两个山谷的中间线上出现的这个三角形结构，面向José Esteves谷和福尔诺谷——第三面朝向杜罗河和Côa河的交叉处。

这种物质性让人想起当地的石头庭院，并且反映出两种不同的属性：混凝土材质和当地石块的质地与颜色。就建筑表达而言，我们决定使用带有页岩颜色（当地石材）的混凝土——其纹理由几种模具制成，这些模具是直接用现有的岩石制成的。

Côa Valley Museum of Art and Archaeology

To build the Museum of Art and Archaeology in the Côa Valley, we connected different parameters: topography, accessibility and program. The fusion of these aspects was highly important for the concept definition – to conceive a museum as an installation on the landscape.

The topography was crucial in our decisions since we had to deal with an enormous difference between the museum entrance and the interior of the building. The arrival is made at the top of this topography and that was one of the reasons to create a platform, a terrace, a belvedere – the stage scenario is the huge and impressive landscape with mountains and valleys. The arrival, a contemplation moment and a parking space as well – is at the same time the museum entrance. The body strategy, in relation to topography, is natural; this means that the platform assumes the same level of the street that ends on the building while the terrain goes down along the side walls showing, in the front elevation, the total height of the volume.

On the higher level of the terrain, the monolithic triangular form is a direct result of the valley's confluences. This triangular body happens in the middle line of the two valleys, facing them (Vale de José Esteves and Vale do Forno) – the third side faces the crossing of Douro and Côa rivers.

Its materiality evokes the local stone yards and reflects two different natures: the concrete's matter and the local stone's texture and color. For the building expression we decided to use concrete made with shale pigment(local stone) – the concrete's texture was made with several molds that were extracted directly from the existing rocks. Camilo Rebelo+Tiago Pimentel

DOURO
VILA NOVA DE FOZ CÔA
CÔA

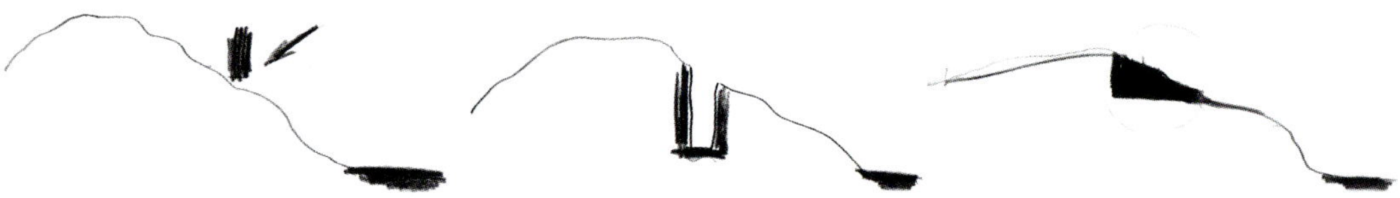

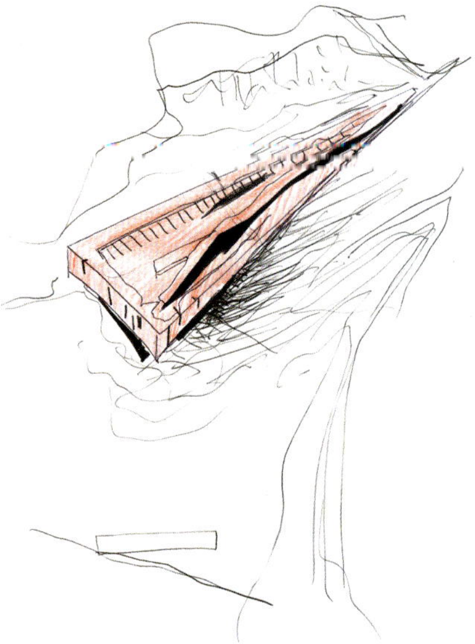

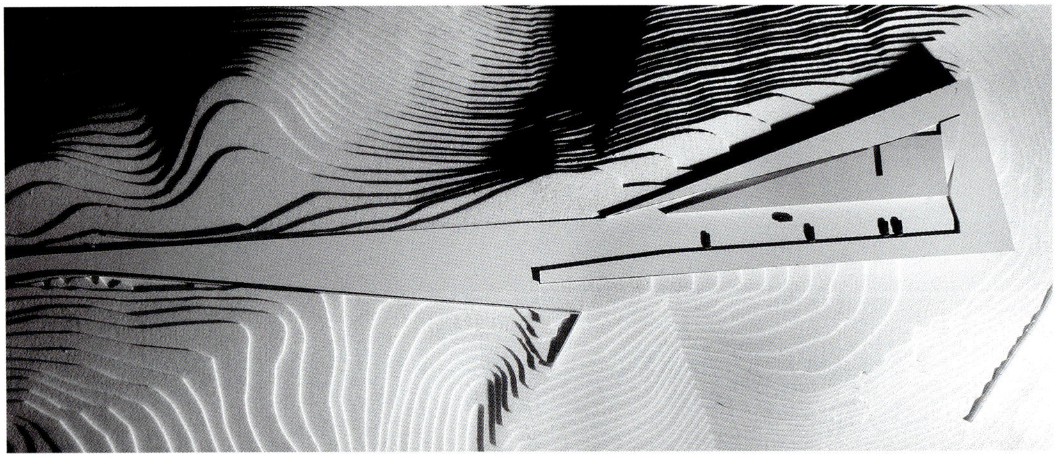

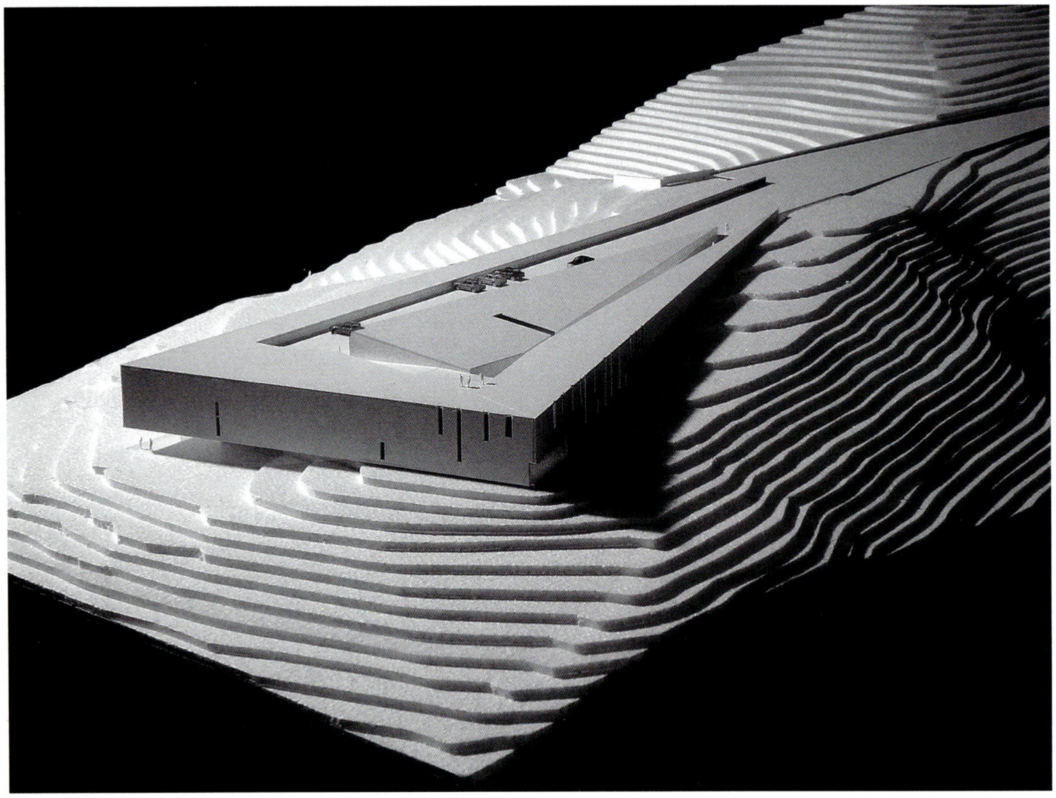

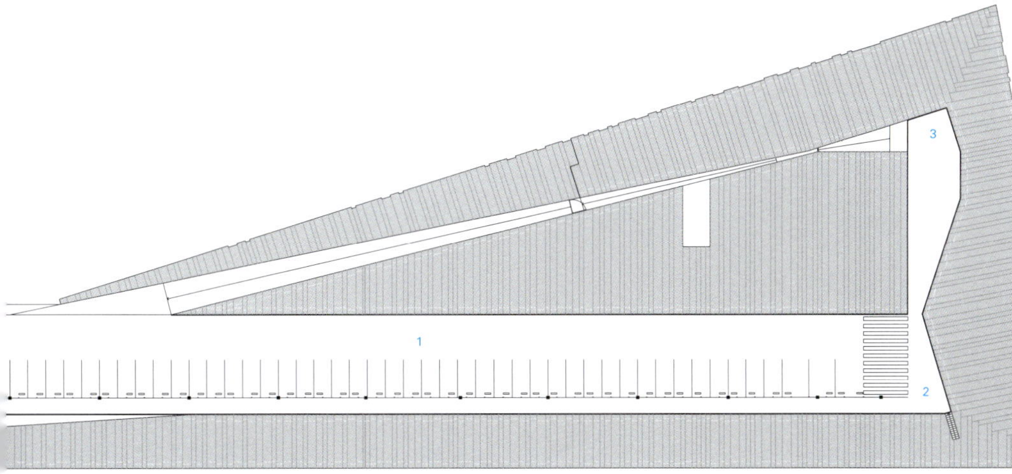

一层
first floor

1. parking area
2. south belvedere access
3. belvedere terrace
4. public access–exterior atrium
5. public access–north entrance
6. education center
7. investigation/archive rooms
8. auditorium entrance
9. auditorium
10. museum–interior atrium
11. bookstore
12. public atrium–elevators access
13. bathrooms
14. storage
15. lockers
16. storage
17. temporary exhibition room
18. exhibition rooms corridor
19. permanent exhibition rooms
20. depot
21. museography room
22. workshop room
23. photography laboratory
24. stair–entrance atrium/restaurant
25. public atrium–restaurant
26. parking space
27. technical area
28. auditorium
29. bathrooms
30. cafeteria
31. bar/restaurant
32. restaurant
33. kitchen
34. storage
35. bathrooms
36. service access
37. service area
38. technical area

地下一层
first floor below ground

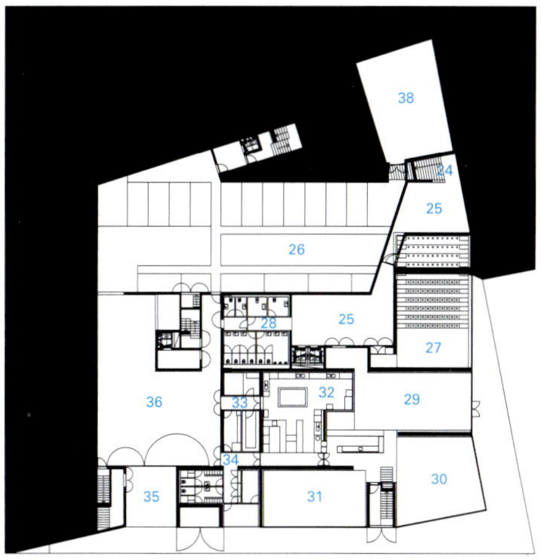

地下二层
second floor below ground

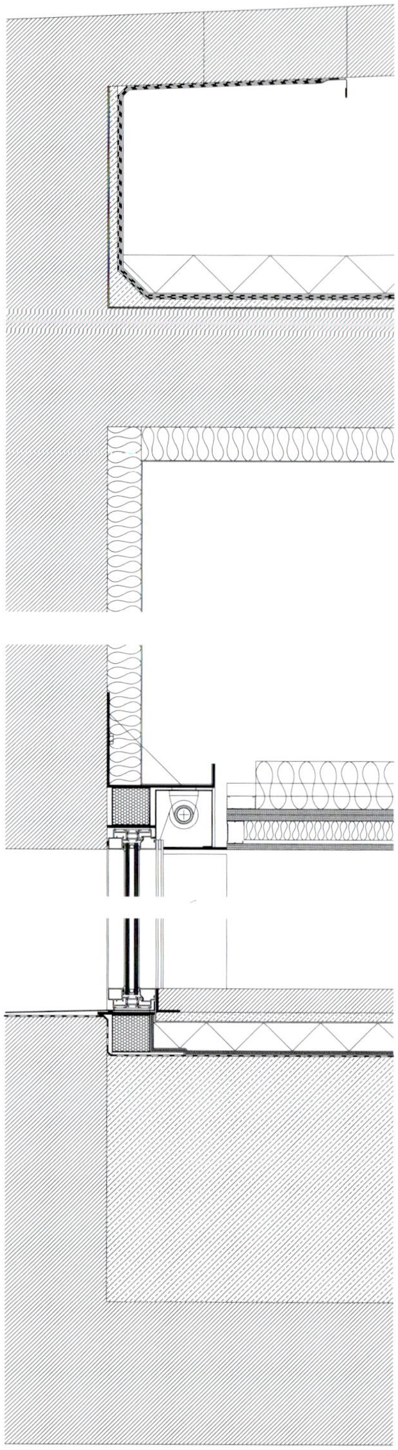

展览室详图
exhibition room detail

钢AISI 316
steel AISI 316

石膏板
plasterboard

保温层——屋顶装置
insulation–roofmate

保温层——墙体装置
insulation–wallmate

保温层——泡沫玻璃
insulation–foam glass

混凝土
concrete

填充物
filling

水泥——饰面层
cement–finishing layer

水泥
cement

砂石
woolstone

砂石
tout-venant

地面
ground

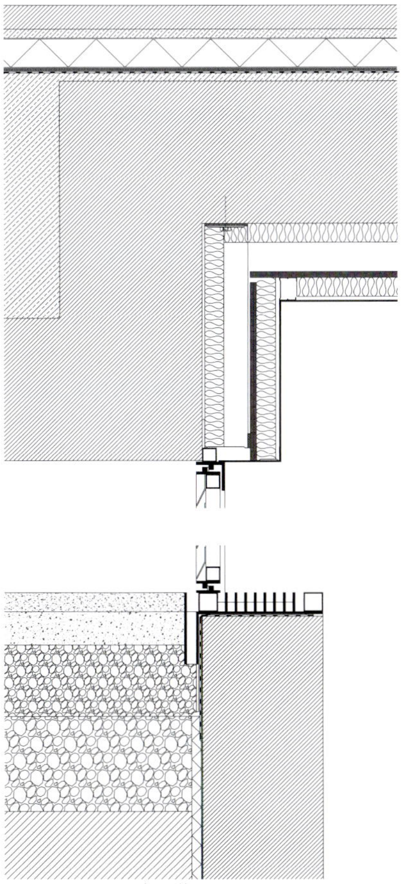

餐厅详图
restaurant detail

项目名称：Museum of Art and Archaeology in the Côa Valley
建筑师：Camilo Rebelo, Tiago Pimentel, Sandra Barbosa
合作者：Bruno Guimarães, Cláudio Reis, Marcelo Correia, Cristina Chioau
工程师：G.O.P. Gabinete de Organização e Projectos
基础和结构：Jorge Nunes da Silva
机械装置和暖通空调：Raul Bessa
电气装置、保安、IT网络：Alexandre Martins
污水处理和水供应：Raquel Fernandes
音响效果：Maria Rosa Sá Ribeiro
地点：Vila Nova de Foz Côa, Guarda ,Portugal
用途：Museum
基地面积：88,294.8m²
建筑面积：6,243.28m²
总建筑面积：8,121.31m²
景观面积：81,882.6m²
建筑覆盖率：7.1 %
楼面比率：9.2 %
建筑规模：three stories above ground
结构：concrete
外观处理：concrete
设计时间：2004~2009
竣工时间：2009
摄影师：courtesy of the architect (©Camilo Rebelo) - p.71, p.73, p.74, p.75, p.77 top
©Nelson Garrido - p.71, p.73, p.77 bottom, p.78~79

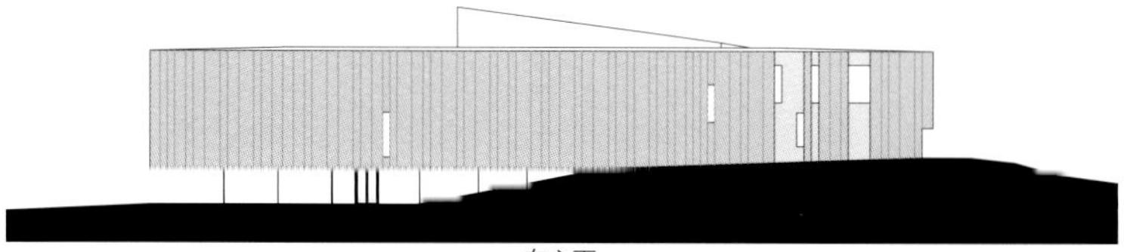
东立面
east elevation

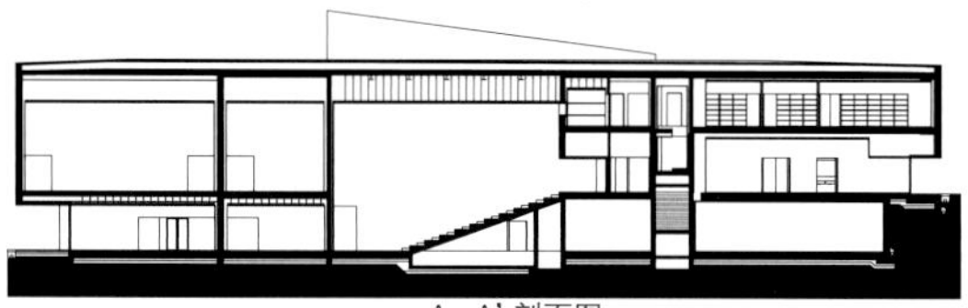
A－A' 剖面图
section A-A'

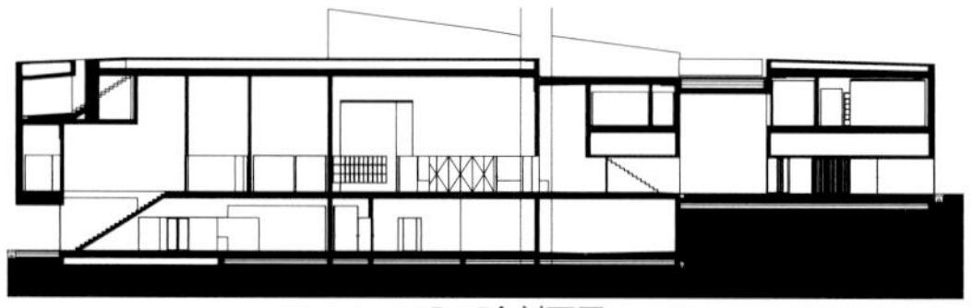
B－B' 剖面图
section B-B'

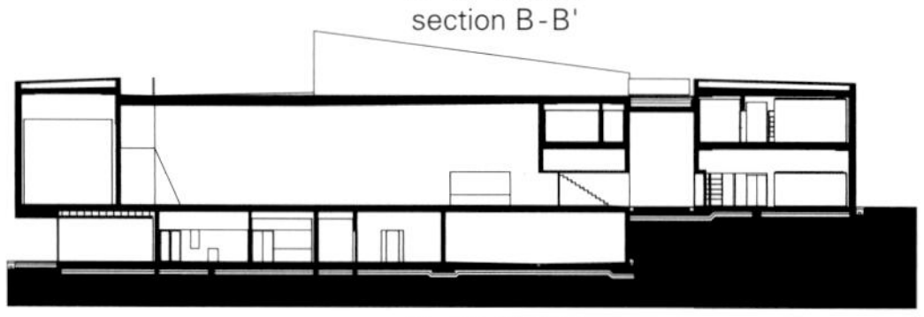
C－C' 剖面图
section C-C'

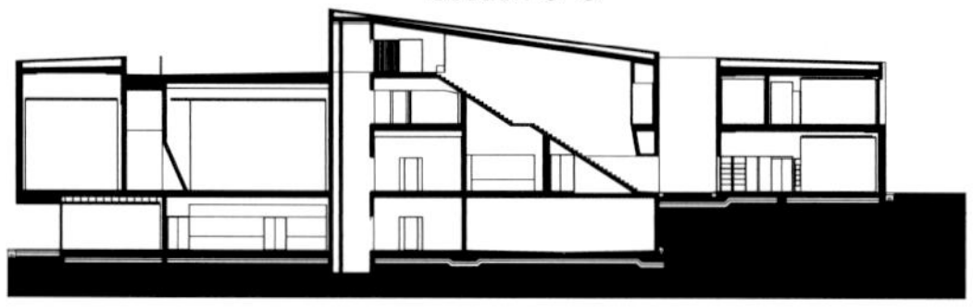
D－D' 剖面图
section D-D'

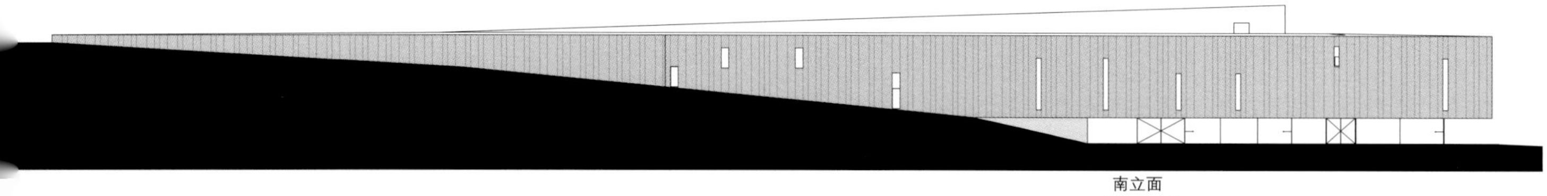
南立面
south elevation

E－E' 剖面图
section E-E'

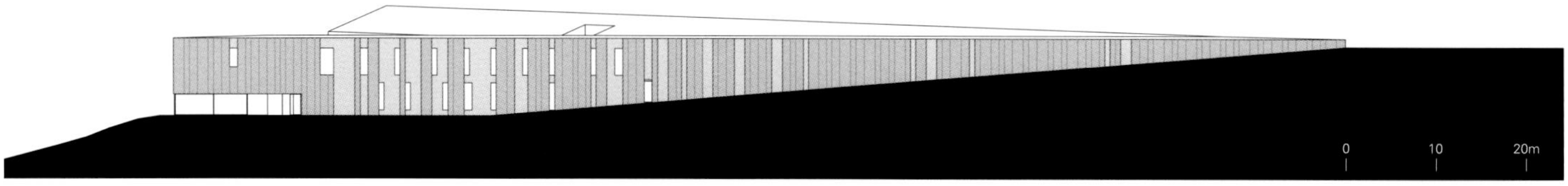

西北立面
north-west elevation

广东省博物馆

Rocco Design Architects

广东省博物馆位于广州珠江新城，是该地区新金融中心的四大文化标志性建筑之一。

这个五层楼博物馆的建筑总面积约为67 000m²。

它被视为一件规模巨大的艺术品、一件雕工精湛的中国古董文物，如漆盒、象牙球、玉碗或青铜罐这些反映时代的珍品的寓言。新博物馆不仅是陈列各种各样珍品的场所，它本身还是一件极具吸引力的珍品，成为一种可识别的文化符号，给游客留下了难忘的旅行经历，使他们感受到当地的历史和传统文化，同时也为增强与欣赏城市文化特性做出贡献。

博物馆的空间布局灵感来源于传说中的同心象牙球雕刻艺术品，每一个雕刻部分都通过盒子展现内部不同的层面和不同的透明度，形成一个有趣的立体格局，通过内部的展示吸引游客。内部空间交织在一起也显示了中庭走廊、各个展厅和后台服务区在视觉和实际上的联结与分离的关系。这种刻意的安排不仅加强了藏宝盒概念的清晰性和连贯性，还显示了整个展览空间在规划和运作上的灵活性。此外，每个主要展厅外墙都呈现出随机的几何动态凹陷。它们是从室内观察室外的视觉突破口，充满自然光线，同时也形成展厅之间的过渡，在那里游客可以享受亲切、均衡协调的休息空间。

主要立面的整体处理方式也基于对象牙球的模仿。使用的材料有铝板、多孔玻璃和玻璃纤维增强水泥板，每个立面都带有设计独特的、各式各样的几何凹陷空隙。为使博物馆与邻近的景色顺利地过渡，设计师在升高的博物馆下方引入了一个起伏的景观层，象征着一块打开极其珍贵的艺术品的红绸布。

Guangdong Museum

The Guangdong Museum is one of four major cultural landmark buildings for the new financial hub in Zhujiang Xincheng (Pearl River New Town) of Guangzhou.

The five-story museum has a total floor area of approximately 67,000 square meters.

It is conceived as an Objet d'Art at a monumental scale, an allegory to the impeccably and intricately sculpted antique Chinese artefacts, such as a lacquer box, an ivory ball, a jade bowl or a bronze tank, which collects and reflects treasures of the times. The new museum is not only designed to house a great variety of fascinating objects of treasure, it is also in itself designed as a treasured object of great fascination that contemplates to become an identifiable cultural icon, giving the visitors a memorable tour and experience of the local provincial history and traditional wisdom as well as contributing to the appreciation and enhancement of cultural identity of the city.

The Museum's spatial arrangement takes its inspiration from

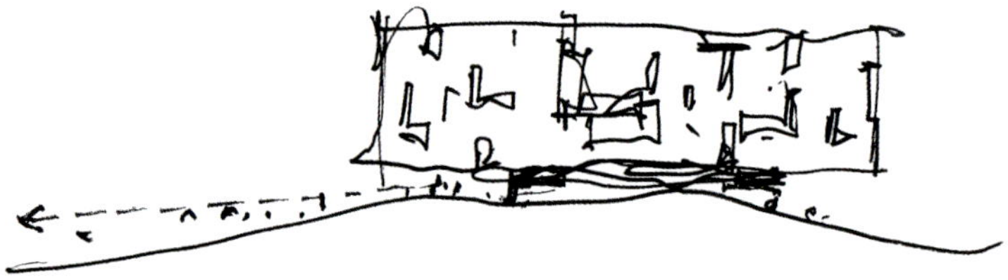

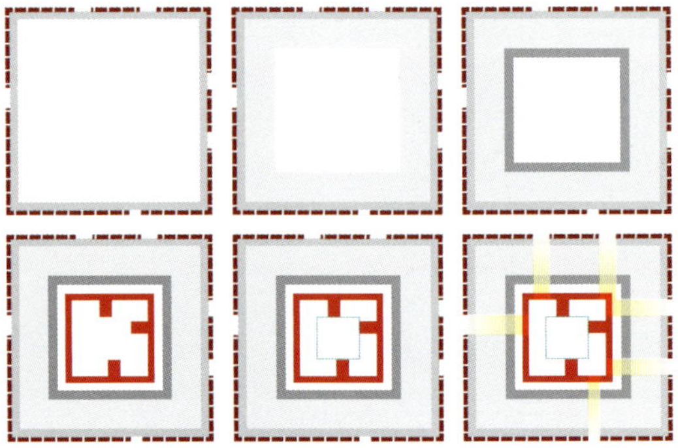

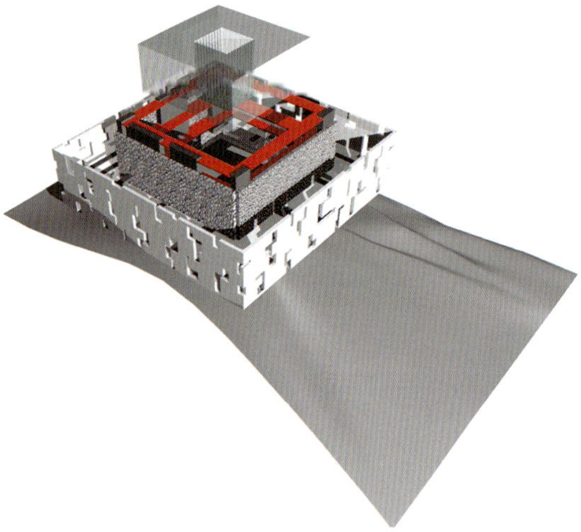

滨江绿化带 THE RIVERSIDE GREENBELT
临江大道 LINJIANG BOVLEVARD

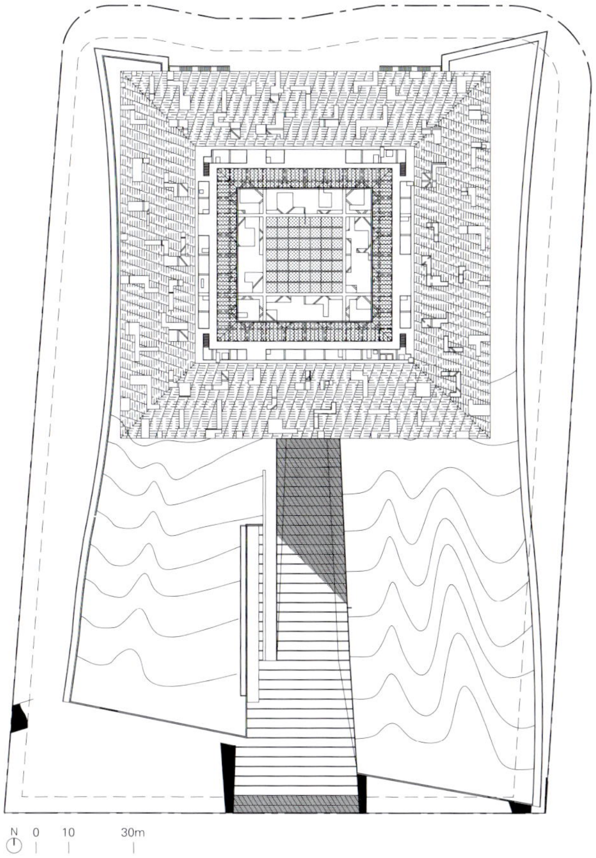
N
0
10
30m

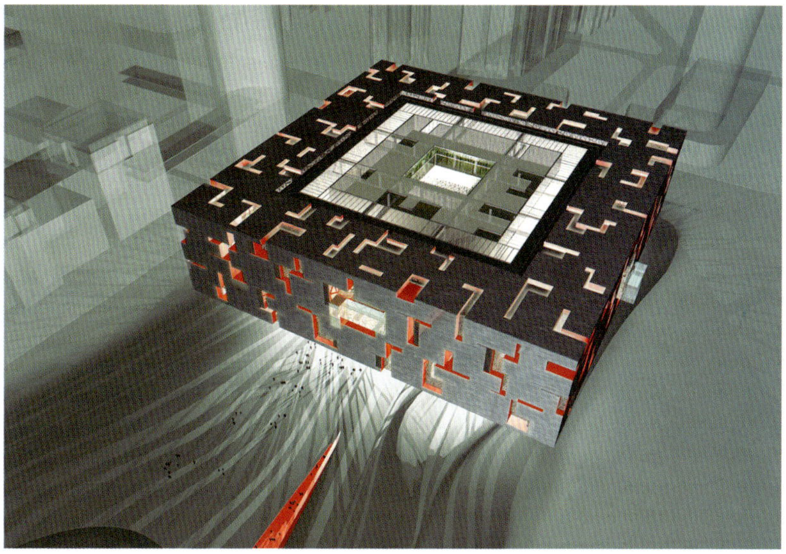

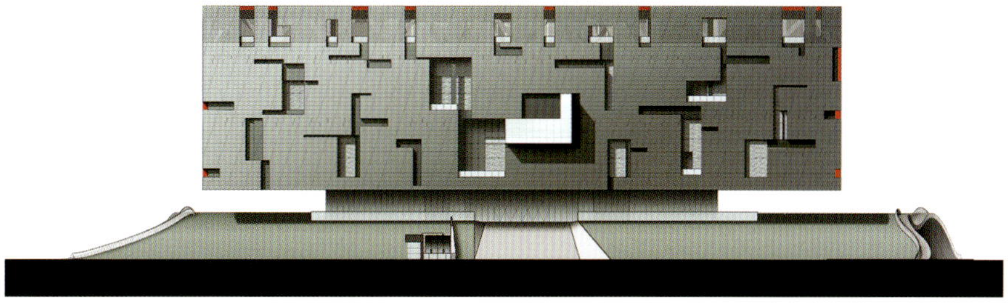

西立面 west elevation

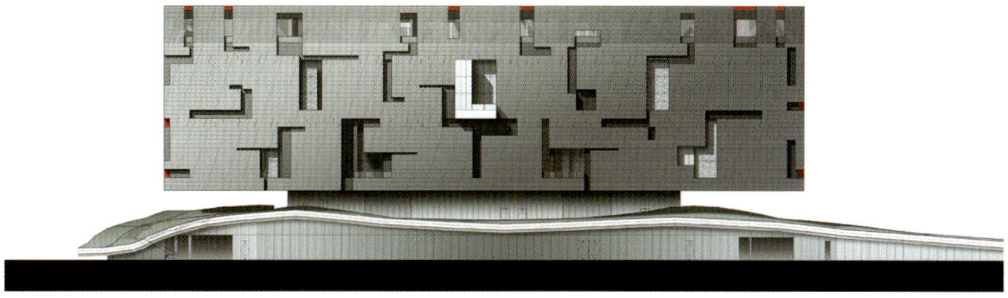

北立面 north elevation

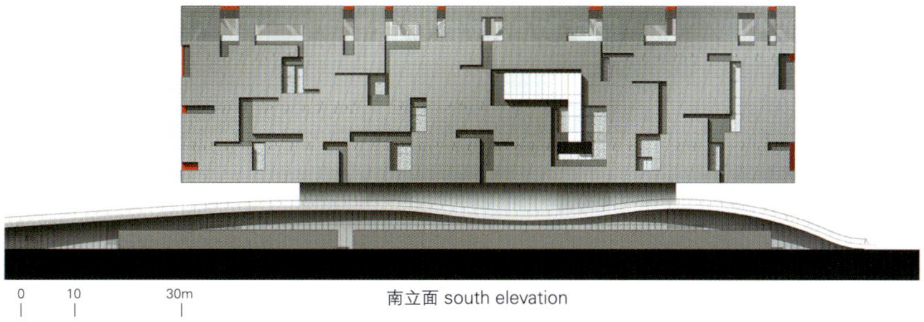

南立面 south elevation

2

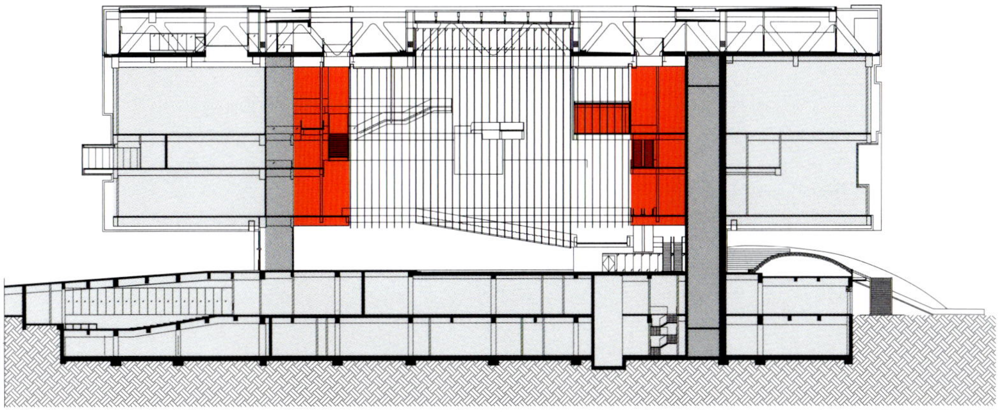

A–A' 剖面图
section A-A'

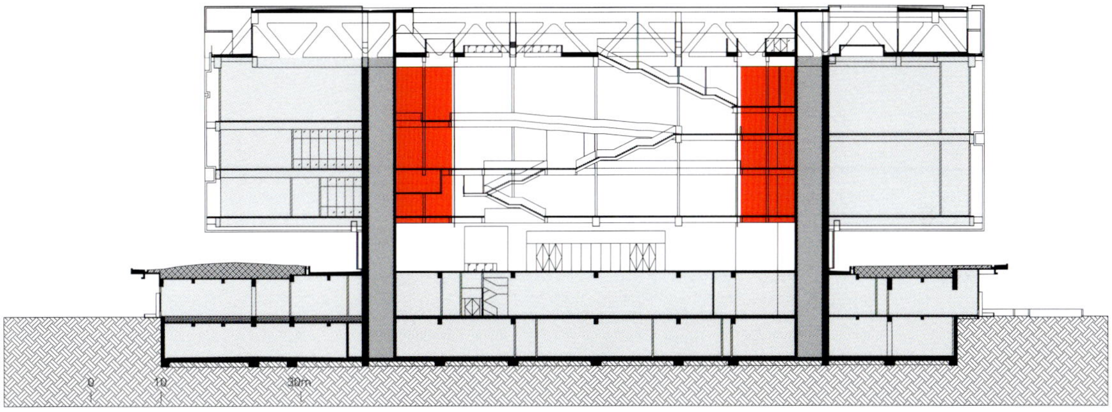

B–B' 剖面图
section B-B'

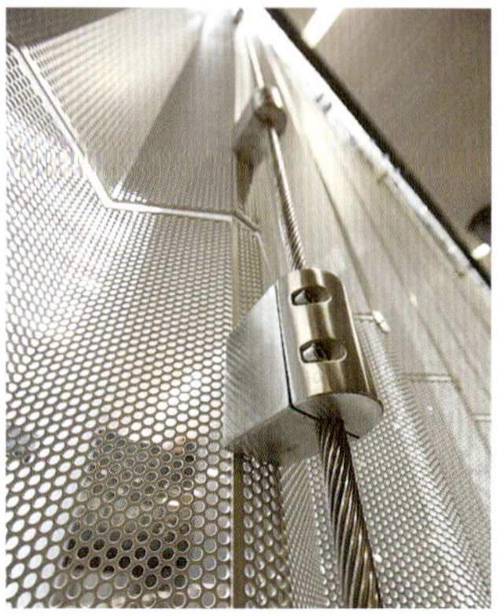

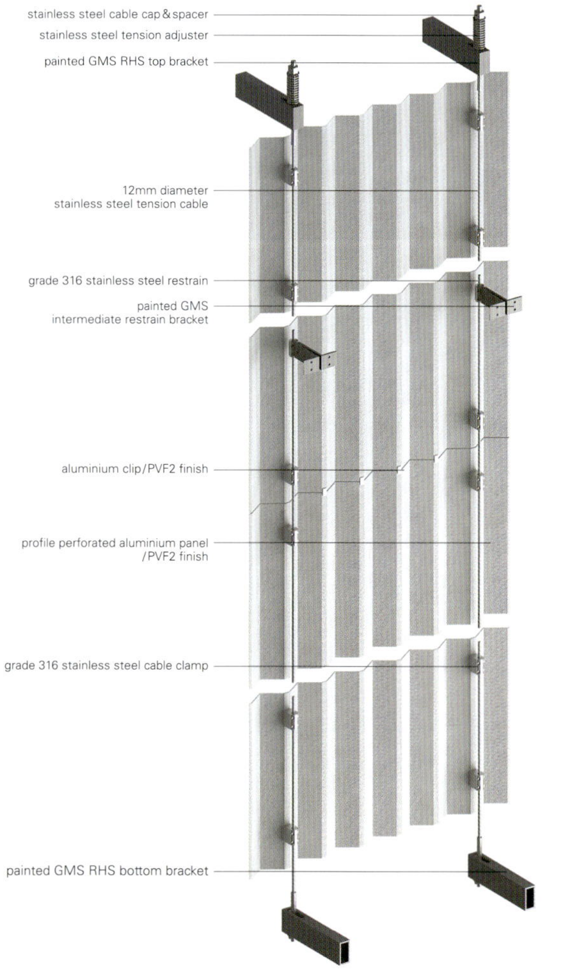

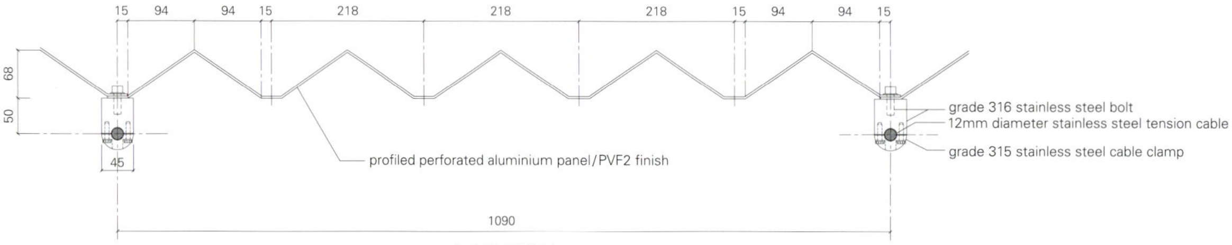

悬垂铝板详图
aluminium hanging screen detail

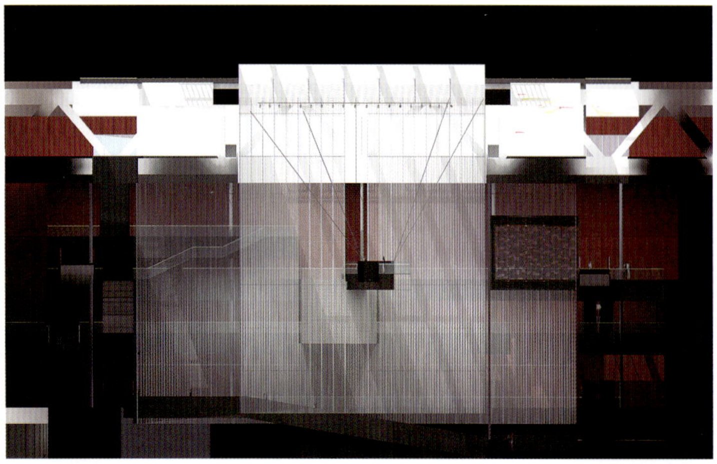

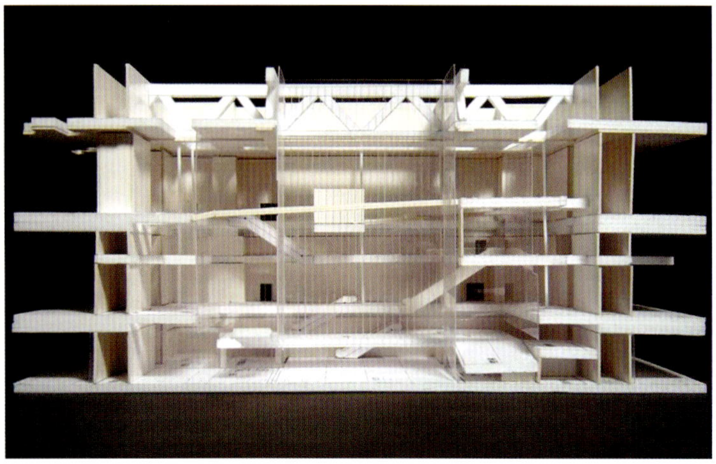

项目名称：Guangdong Museum
建筑师：Rocco Design Architects Ltd
项目团队：Rocco Yim, William Tam, C. M. Chan, Derrick Tsang, Christopher Wong, Ricky Wang, Lucia Cheung, Grace Lin, Ivan Chui, Christina Chan, Amber Wang
景观建筑师：Rocco Design Limited in association with The Architectural Design and Research Institute of Guangdong Province
当地设计院：The Architectural Design and Research Institute of Guangdong Province
结构设计顾问：Ove Arup & Partners Hong Kong Ltd
立面顾问：Ove Arup & Partners Hong Kong Ltd
声学顾问：Ove Arup & Partners Hong Kong Ltd
照明顾问：Light & View Design
甲方：Guangdong Museum
地点：Zhujiang Xincheng (Pearl River New Town), Guangzhou, China
基地面积：41,000m²
总建筑面积：67,000m²
竣工时间：2010
造价：RMB 884M
摄影师：courtesy of the architect-p.85, p.88, p.89, p.91
Courtesy of the architect (©Almond Chu)-p.83, p.84, p.87, p.90
Courtesy of the architect (©Marcel Lam)-p.80~81, p.86

the legendary concentric ivory balls carving. Each carving slice through the box reveals different layers and varying degrees of transparency within the interiors, forming interesting spatial patterns and luring visitors through its exhibits inside. The interweaving of interior space pockets also reveal the intricate relationship between the visual and physical connection and separation of the atrium corridor, the individual exhibition halls and the back-of-house service areas. This deliberate arrangement not only reinforces the clarity and coherence of the treasure-box concept, but also allows flexibility in planning and operation of all the exhibition spaces. In addition, each of the main exhibition halls are punctured with random alcoves of dynamic spatial geometry. Filled with natural light and served as visual breakouts to the outside, they are also transitions between the exhibition halls which offer visitors intimate and well-balanced resting spaces.

The overall treatment of the main facade is also based on the analogy of ivory ball. Using materials such as aluminum panels, fritted glass and GRC panels, each elevation is uniquely designed with different geometric voids recessed into the building mass. In order to achieve a smooth transition between the museum and the adjoining landscape, an undulating landscape deck is introduced underneath the elevated museum box, metaphorically symbolizing a silk cloth unwrapping a much treasured piece of artwork.

Rocco Design Architects

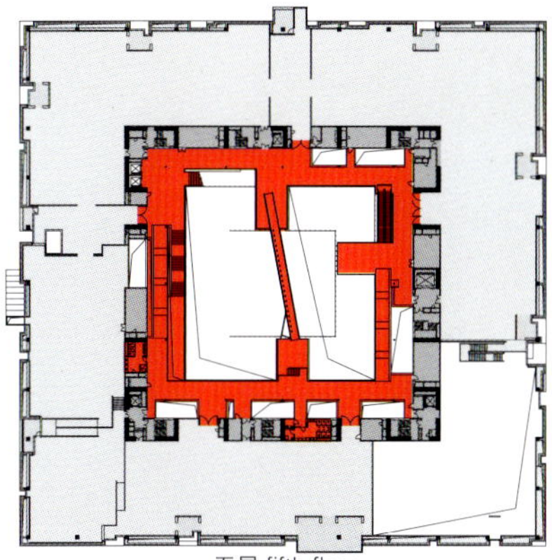

五层 fifth floor

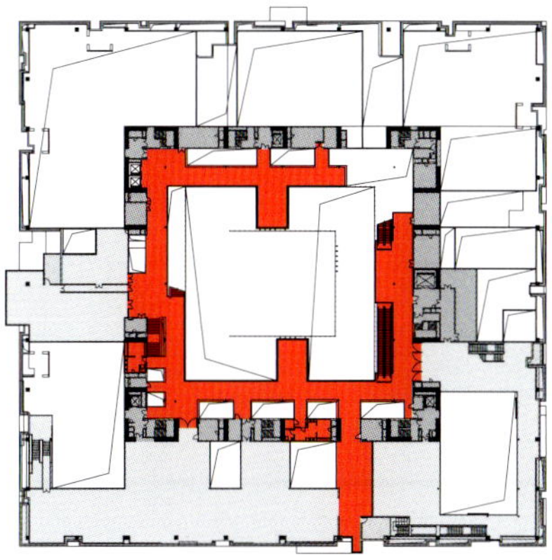

四层 fourth floor

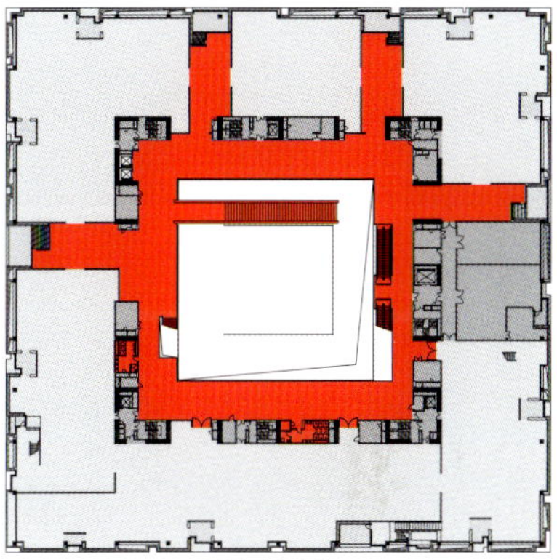

三层 third floor

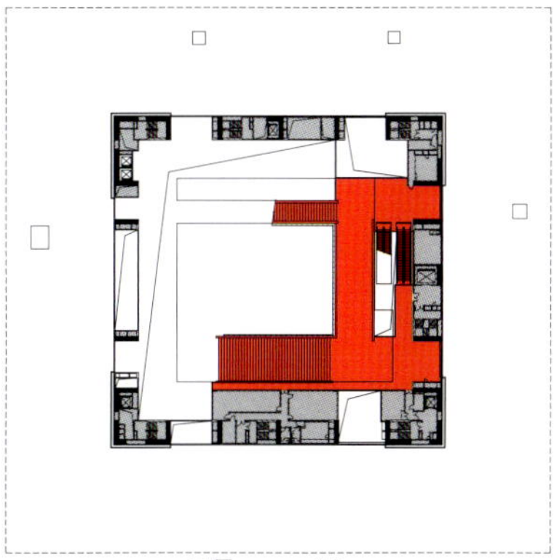

二层 second floor

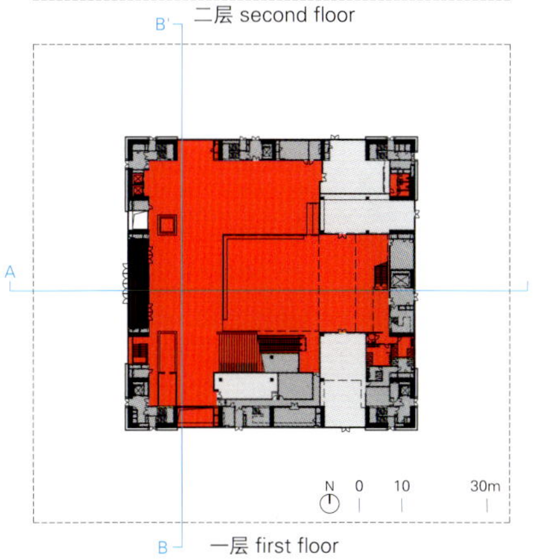

一层 first floor

索玛雅博物馆

FREE - Fernando Romero

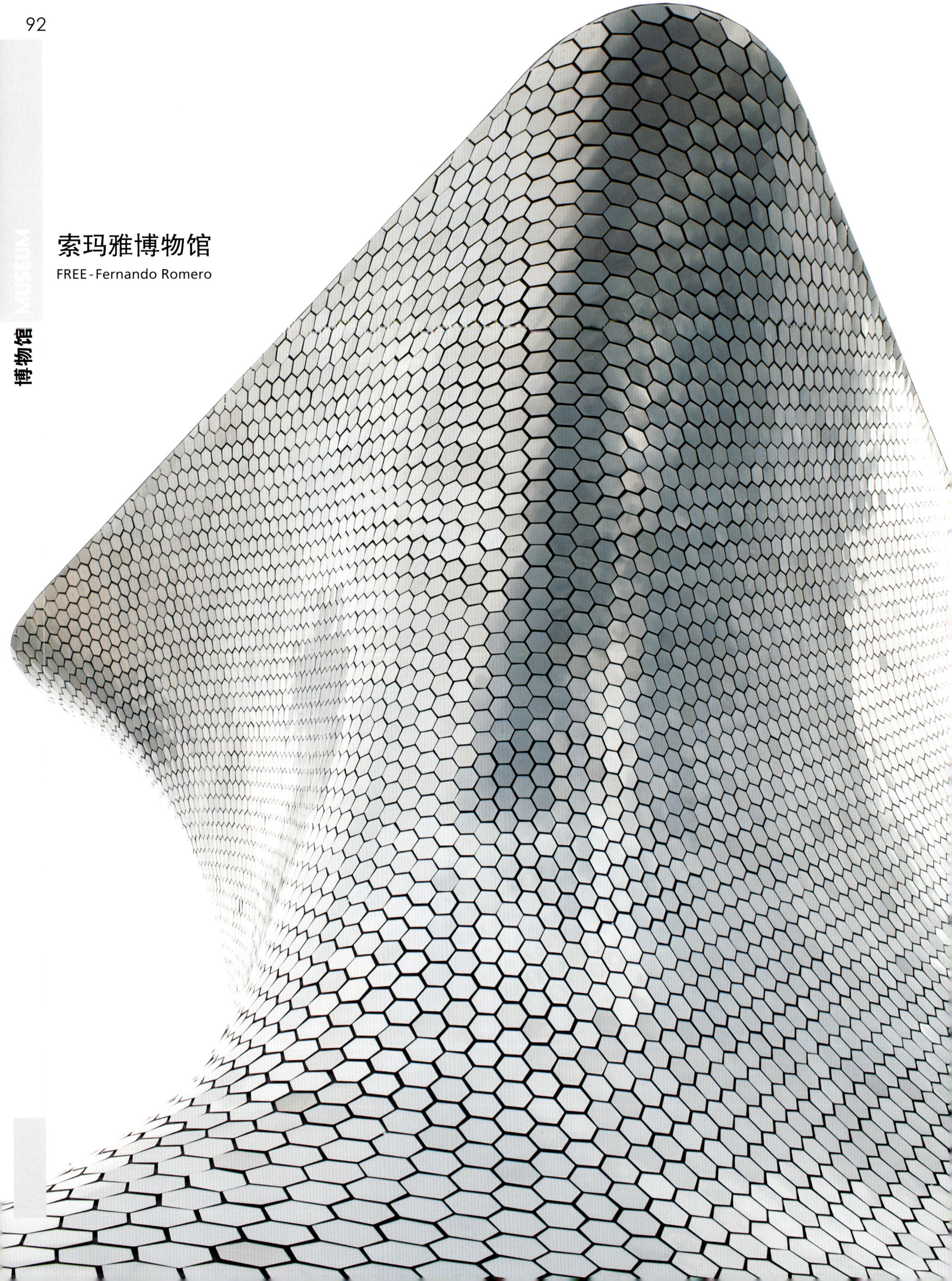

索玛雅博物馆是波朗科区(墨西哥城最繁荣的区域之一)边缘大规模混合用地上都市发展计划的一部分。索玛雅博物馆位于原有的20世纪40年代工业用地的旧址上。作为该地区优秀的文化项目,它在激励该地区都市感知及都市生活转型中发挥着关键作用。这个机构为人们展示都市的混合设计规划,在促进邻近地区商业活动的同时,活跃并加强了周边公共场所的活动。

为了给该场地打造一个崭新的形象,计划需要创造出强烈的都市感和标志性。因此,建筑师们将索玛雅博物馆构想成一座旋转的雕塑建筑,既是墨西哥城的一个建筑物,又是该城市的一部分。博物馆标新立异的外形在墨西哥及国际建筑史上树立了新的典范。

这座建筑的外壳由28根形状及大小各异的独特的弯曲钢柱构成。整个结构由每层楼共七个吊环组成的系统固定。

为给其收藏品创造一个庇护所,建筑不透明的立面由六边形铝块构成,使整个建筑在防腐性及耐用性最佳的情况下将表面开口减到最小。单个铝块的组合体反映出博物馆收藏品的多样性,其中包括世界第二大的罗丹雕塑藏品区,中世纪、文艺复兴时期艺术展区,和一条印象主义藏品画廊。

多样化的藏品放置在6000m²的连续展厅内,而展厅遍布建筑的六层楼。该建筑也包含一系列公私设施,包括:可纳350人的礼堂、图书馆、餐厅、礼品店、多功能休息室、行政办公室。非线性交通流线区域将这些设施连接起来,这样当用户沿着一条普通小路穿梭于建筑中时,它能为其创造出相互交流的区域,并最终到达顶楼自然采光的多功能画廊。

Soumaya Museum

The Soumaya Museum is part of a large scale mixed-use urban development on the edge of the Polanco district, one of the most exclusive areas of Mexico City. Located on a former existing 1940's industrial site, the Soumaya Museum – as the pre-eminent cultural program in the area – is a key instigator towards the transformation of urban perception and urban life in the area. Its role as an institution provides urban programmatic mixture, activating and intensifying the surrounding public space while promoting the

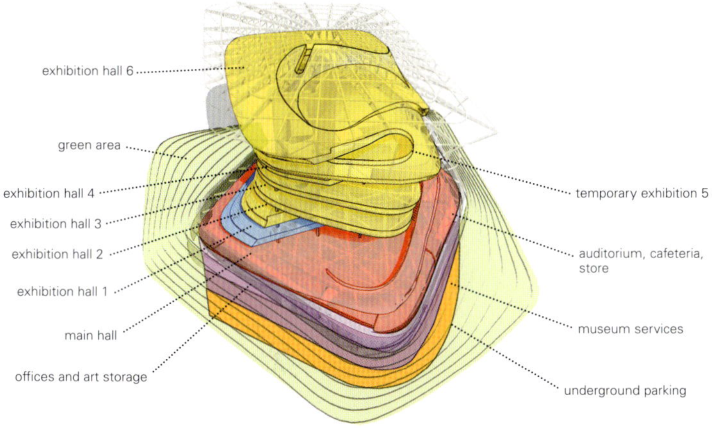

规划分布图
programmatic distribution diagram

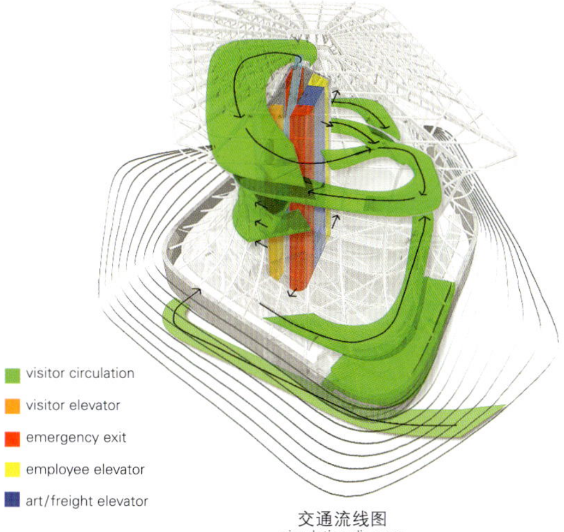

交通流线图
circulation diagram

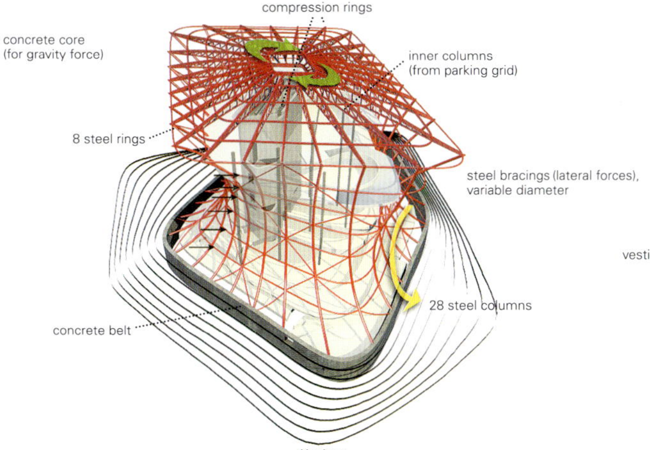

构造图
structural diagram

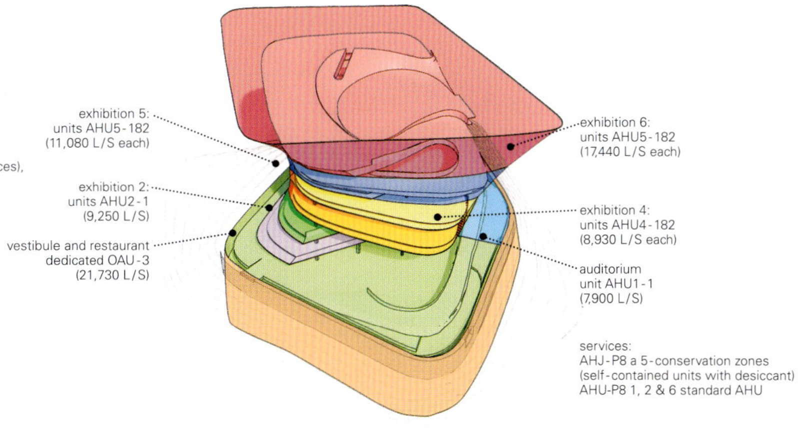

空气调节图
air-conditioning diagram

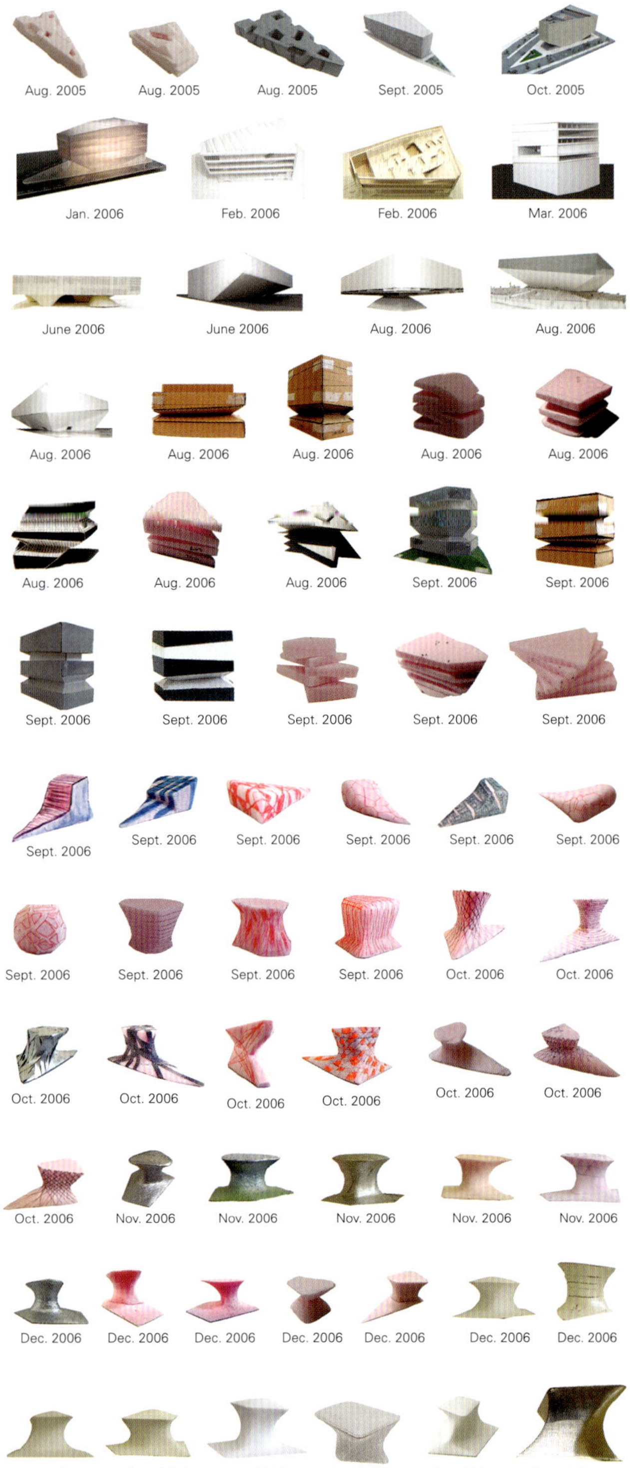
Aug. 2005
Aug. 2005
Aug. 2005
Sept. 2005
Oct. 2005
Jan. 2006
Feb. 2006
Feb. 2006
Mar. 2006
June 2006
June 2006
Aug. 2006
Aug. 2006
Aug. 2006
Aug. 2006
Aug. 2006
Aug. 2006
Aug. 2006
Aug. 2006
Aug. 2006
Aug. 2006
Sept. 2006
Sept. 2006
Sept. 2006
Sept. 2006
Sept. 2006
Sept. 2006
Sept. 2006
Sept. 2006
Sept. 2006
Sept. 2006
Sept. 2006
Sept. 2006
Sept. 2006
Sept. 2006
Sept. 2006
Sept. 2006
Sept. 2006
Oct. 2006
Oct. 2006
Oct. 2006
Oct. 2006
Oct. 2006
Oct. 2006
Oct. 2006
Oct. 2006
Oct. 2006
Nov. 2006
Nov. 2006
Nov. 2006
Nov. 2006
Nov. 2006
Dec. 2006
Dec. 2006
Dec. 2006
Dec. 2006
Dec. 2006
Dec. 2006
Dec. 2006
Feb. 2007
June 2007
June 2007
Aug. 2007
Nov. 2007
Feb. 2008

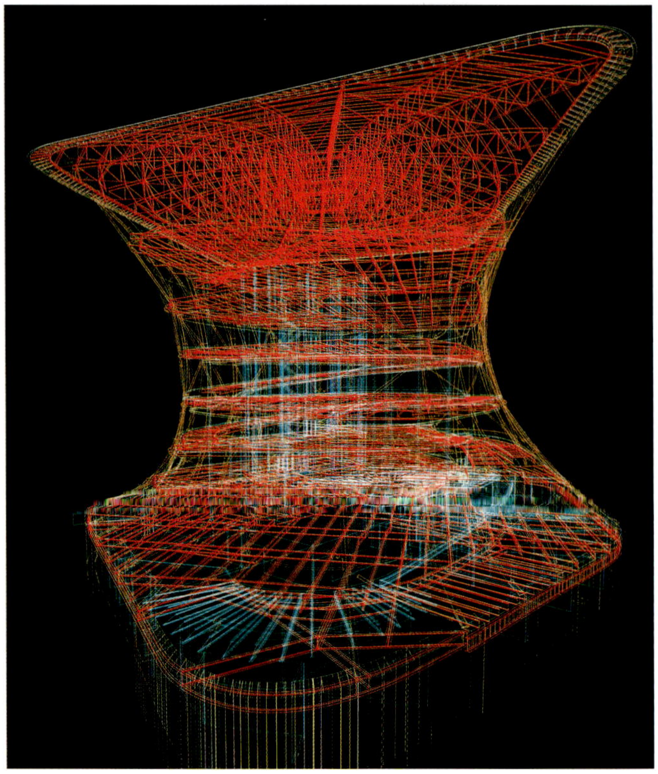

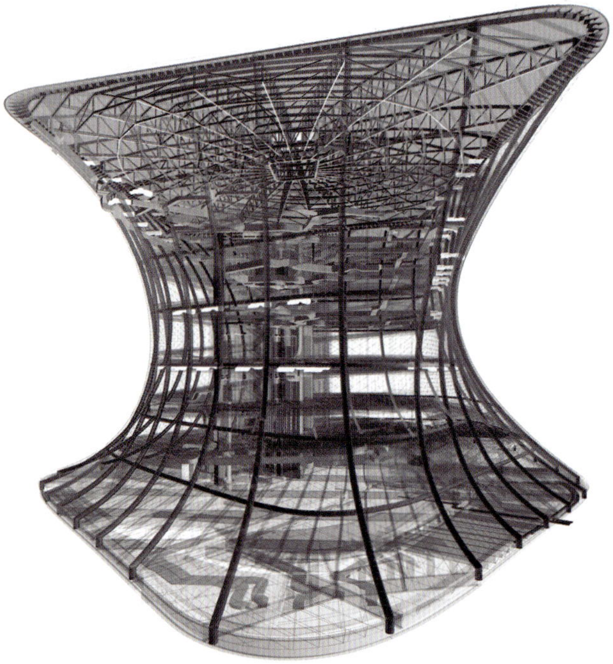

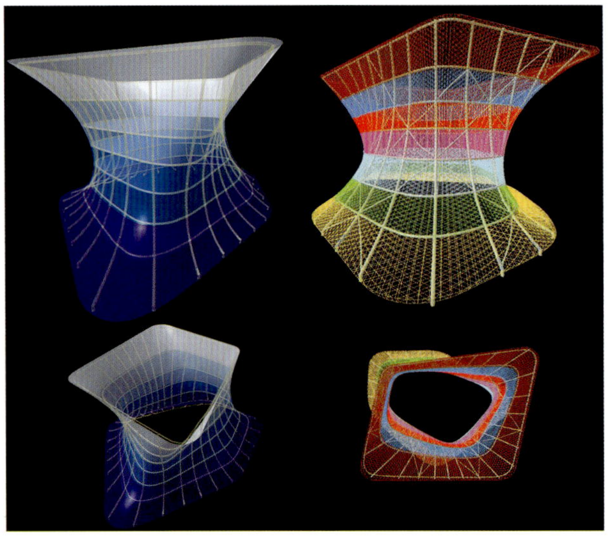

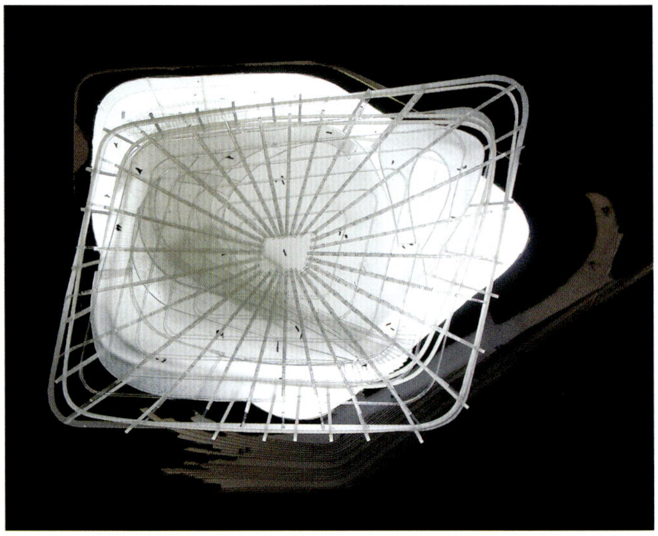

Courtesy of the architect(©Adam Wiseman): left-top, left-third from the top, left-bottom, right-top
Courtesy of the architect(©Ana Paula Herrera): left-second from the top, left-fourth from the top
Courtesy of the architect: bottom-center, bottom-right

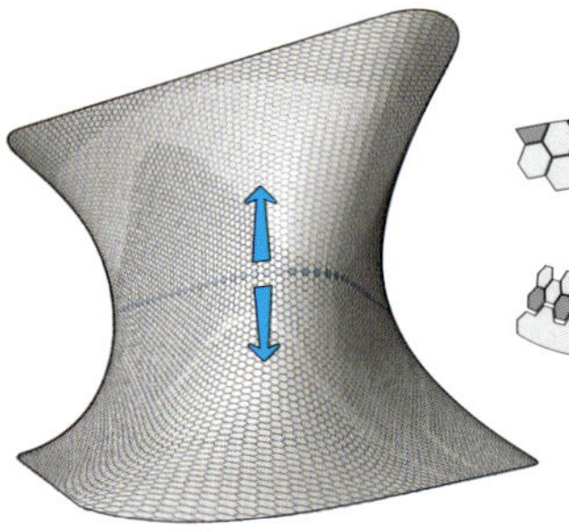

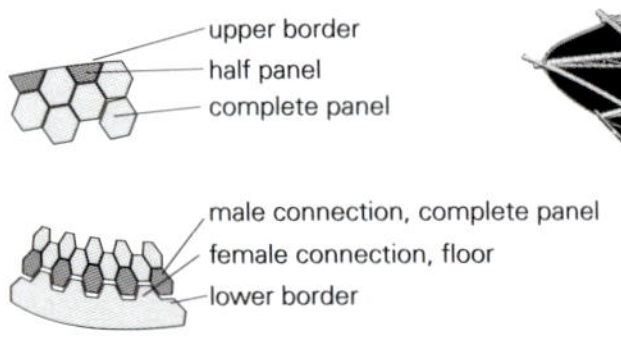

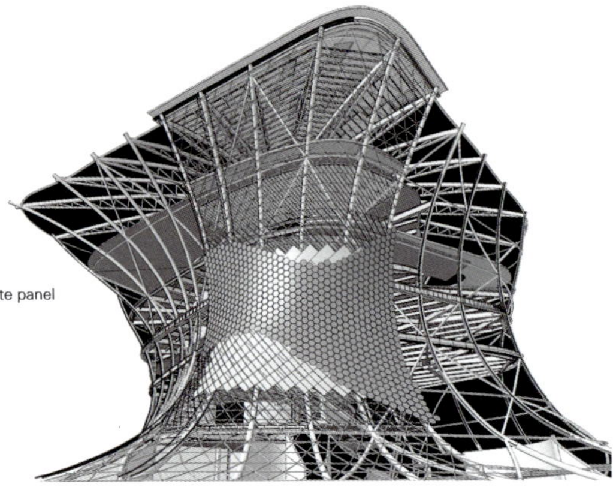

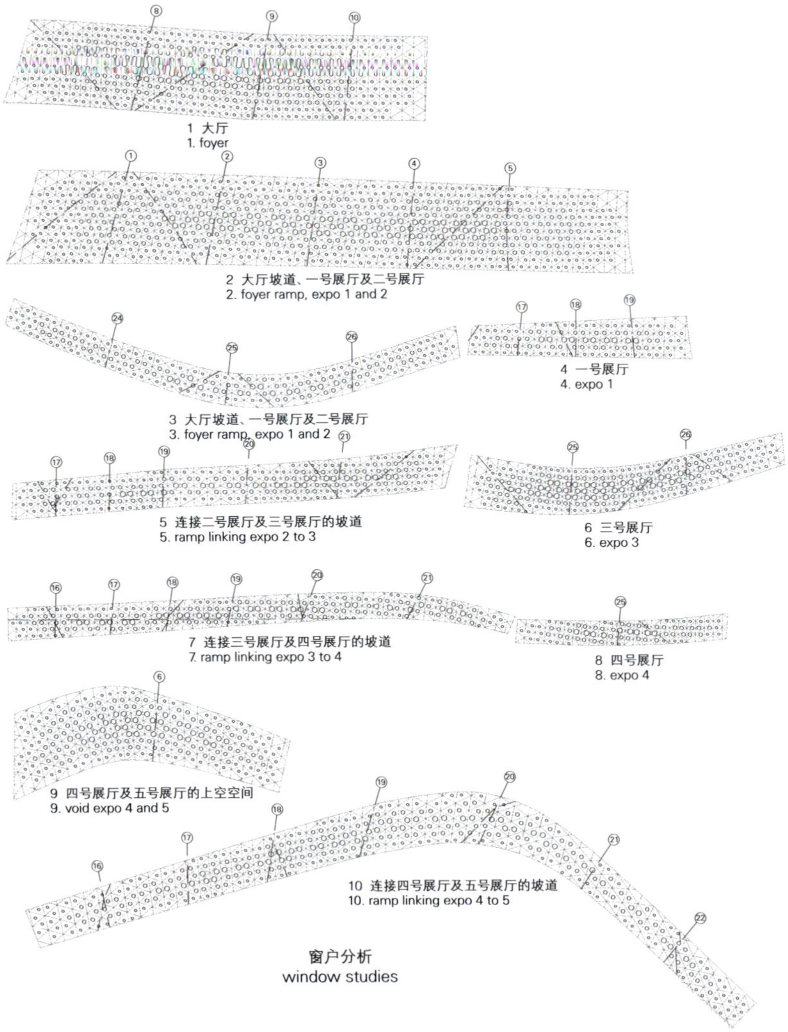

窗户分析
window studies

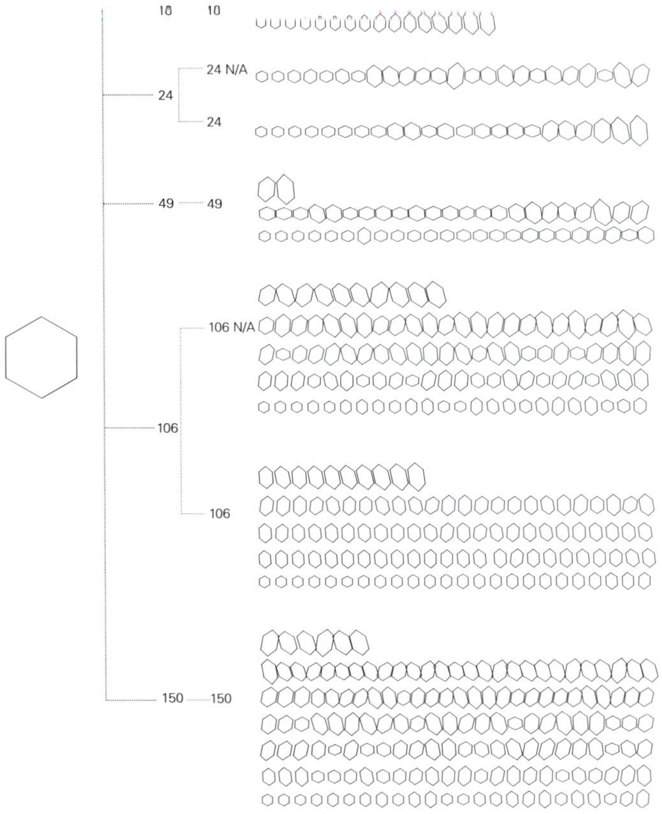

六边形形状分类
hexagon family sorting

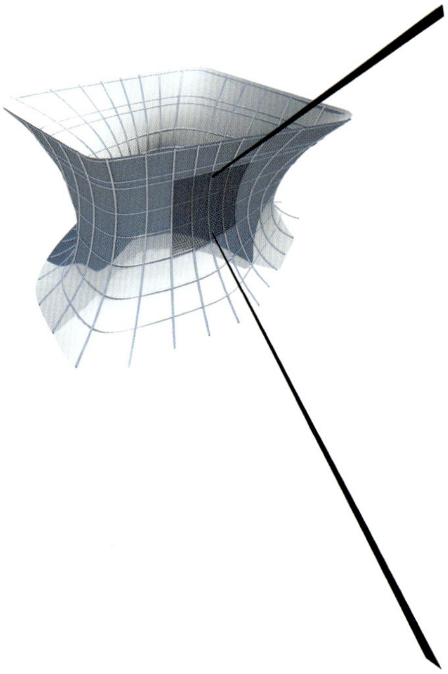

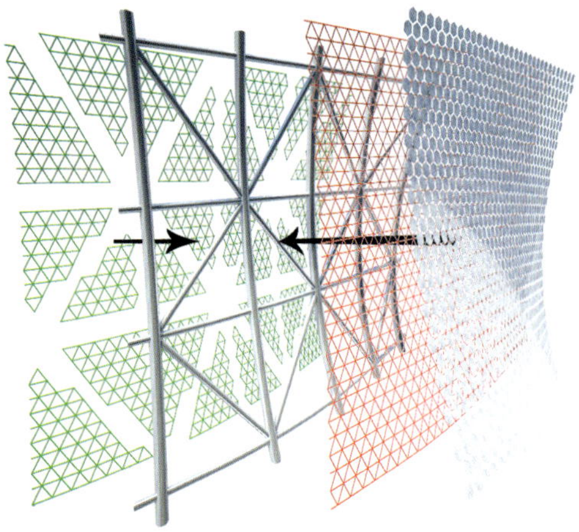

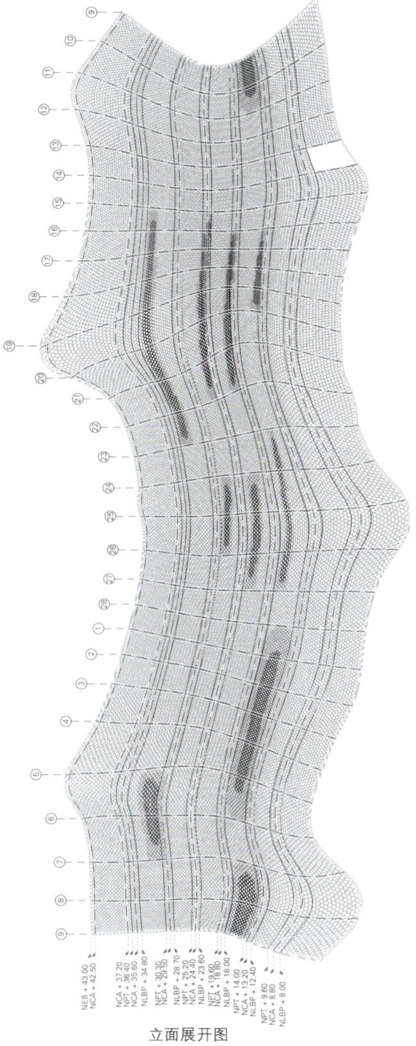

立面展开图
facade unfolded

六边形表面形成过程概念图
hexagon-based surface rationalization process diagram

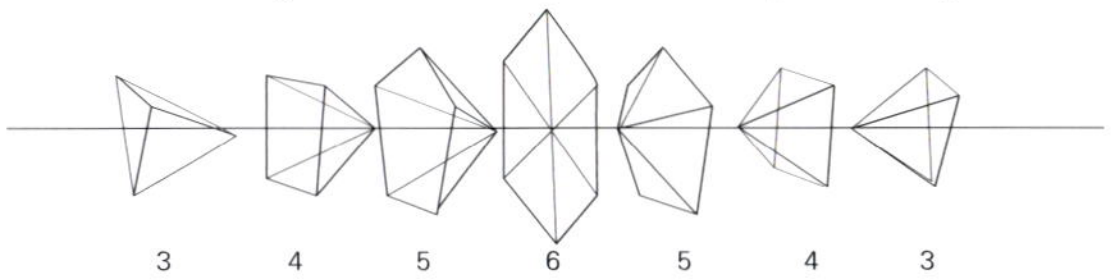

立面几何原理
geometric facade principles

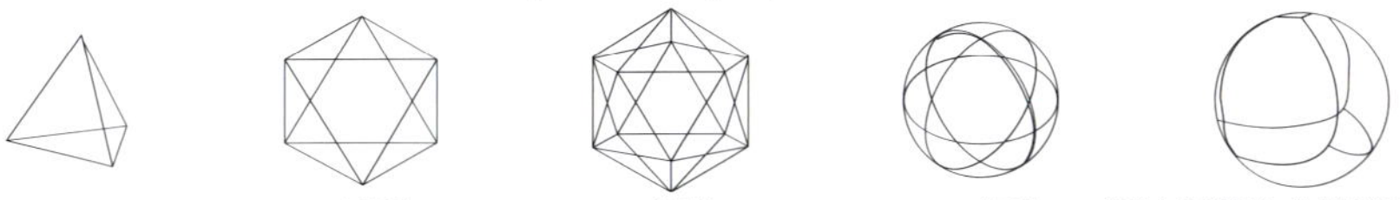

通过六边形置换，分成标准尺寸及定制尺寸的镶板
panel sorting into standards and customs, based on a pure hexagon displacement

球体：由中间开始制作出来的镶板
sphere: panels created from centers

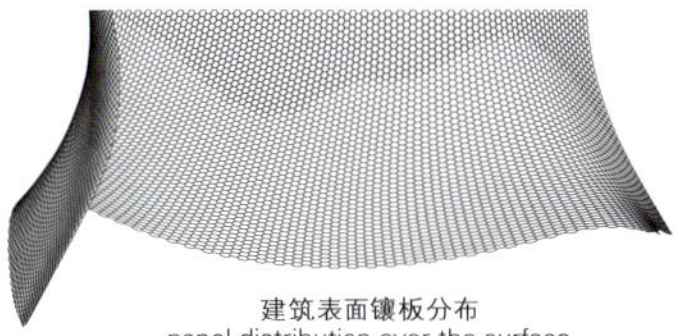

建筑表面镶板分布
panel distribution over the surface

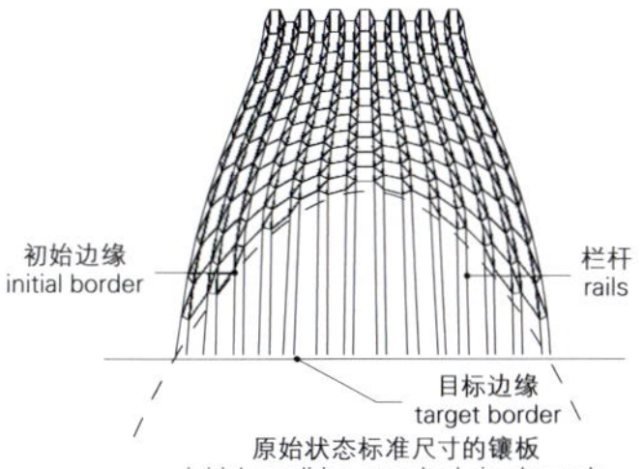

原始状态标准尺寸的镶板
initial condition standard-sized panels

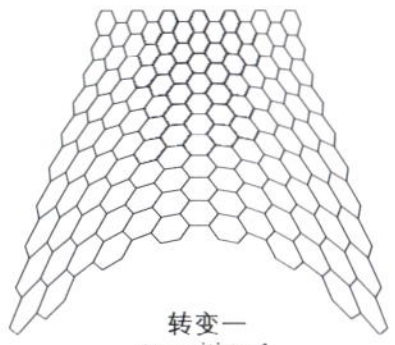

转变一
transition 1

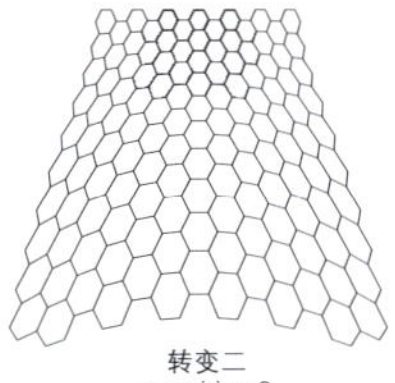

转变二
transition 2

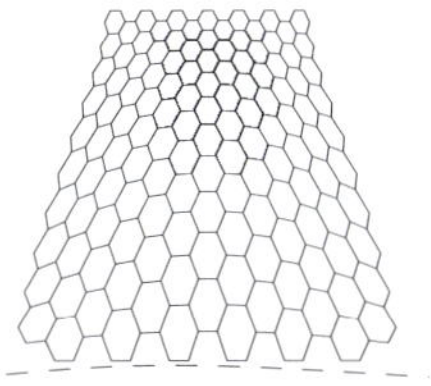

最终状态：标准及定制尺寸的镶板
final condition: standard and custom-sized panels

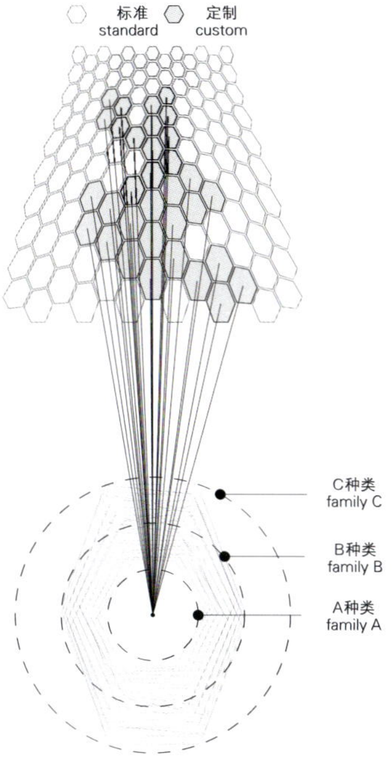

基于与理想的中心之间距离对所有镶板的分类
panel family sorting based on distance from an ideal center

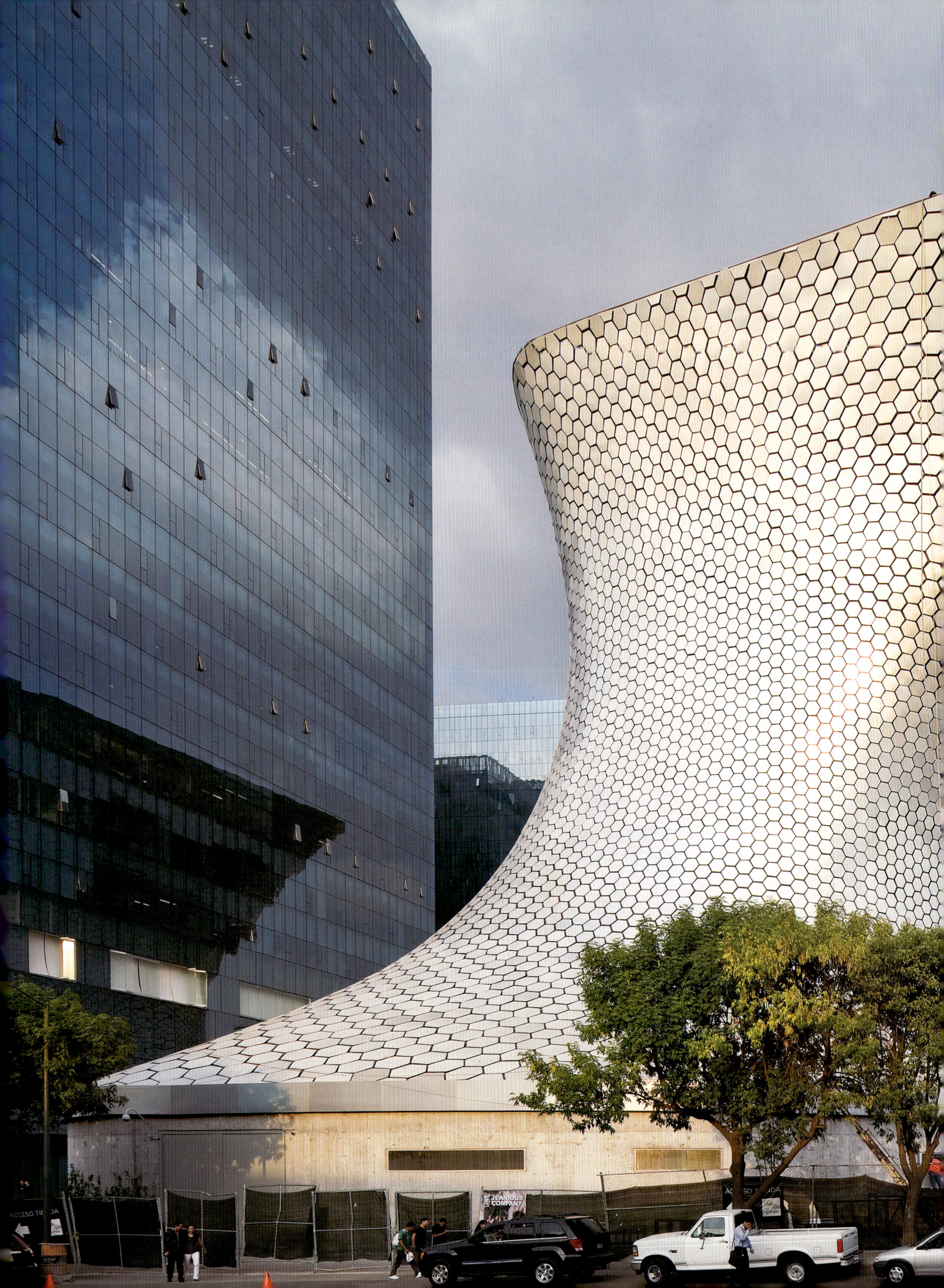
ACCESO TIENDA
JEANIOUS COMPANY
ACCESO TIENDA

Ahora Polanco
ya es VIP
Estimulación
y
Preescolar
¡Te Esperamos!
Inscripciones Abiertas
PREVENTA
SPORT CITY PLAZA CARSO
Aprovecha descuentos

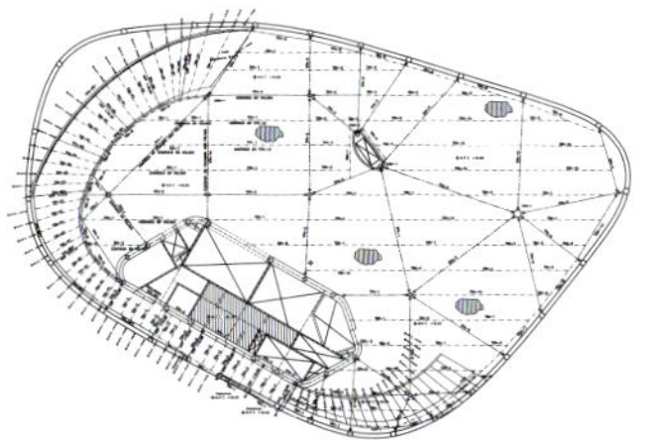

结构楼面标高+19.60m
structural floor level +19.60m

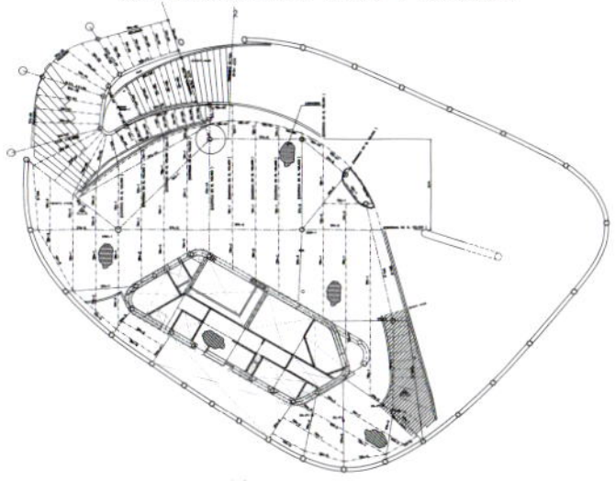

结构楼面标高+14.00m
structural floor level +14.00m

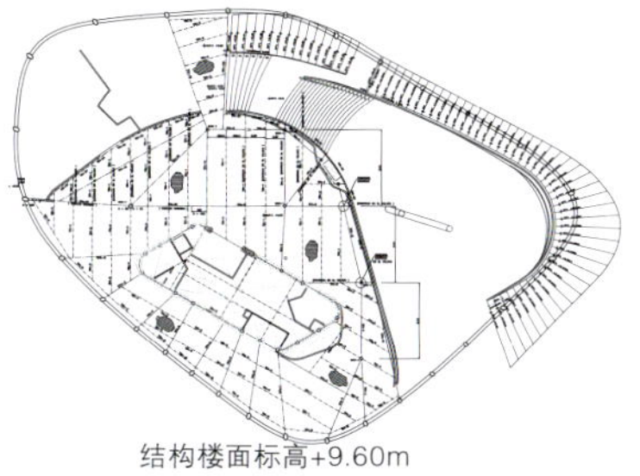

结构楼面标高+9.60m
structural floor level +9.60m

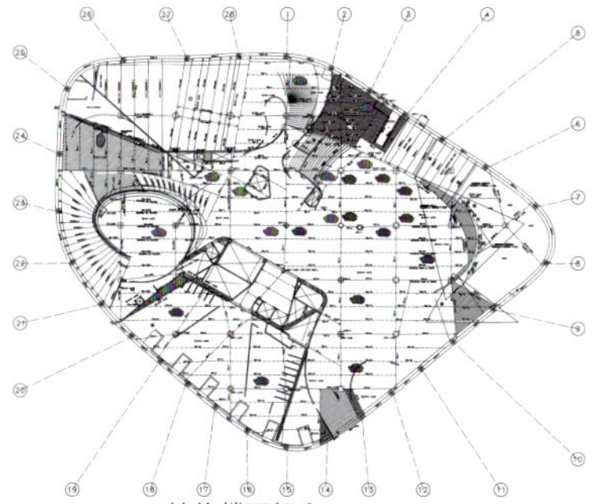

结构楼面标高+5.00m
structural floor level +5.00m

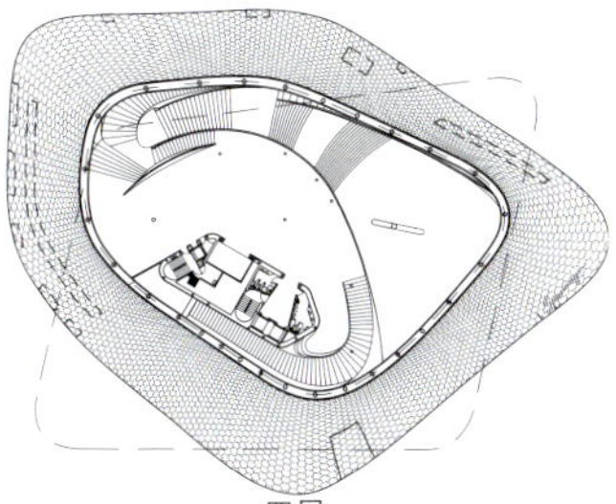

四层
fourth floor

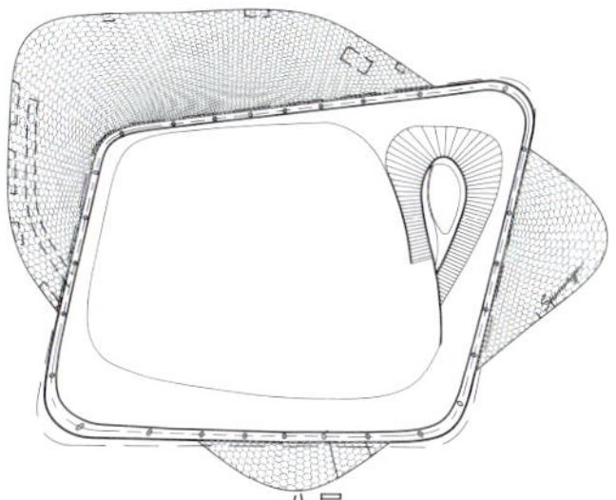

八层
eighth floor

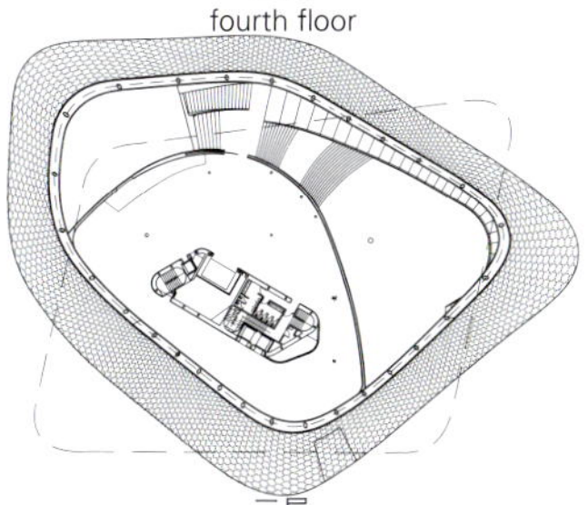

三层
third floor

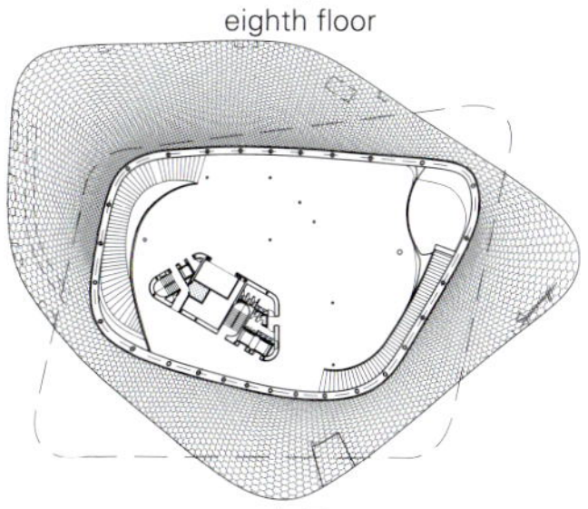

七层
seventh floor

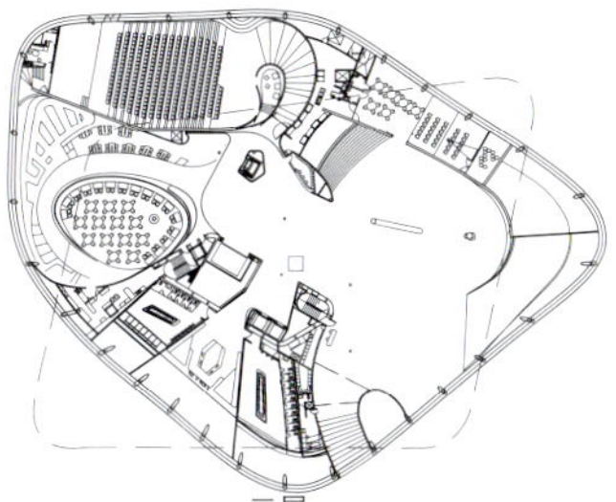

二层
second floor

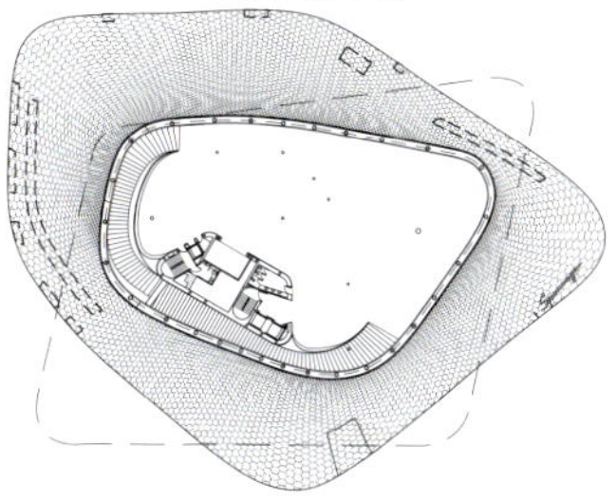

六层
sixth floor

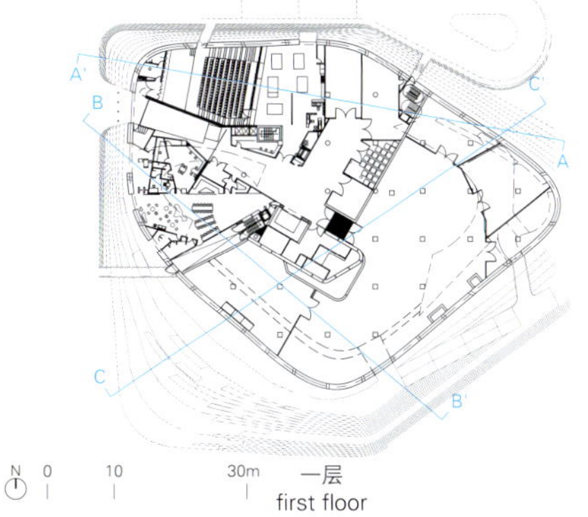

一层
first floor

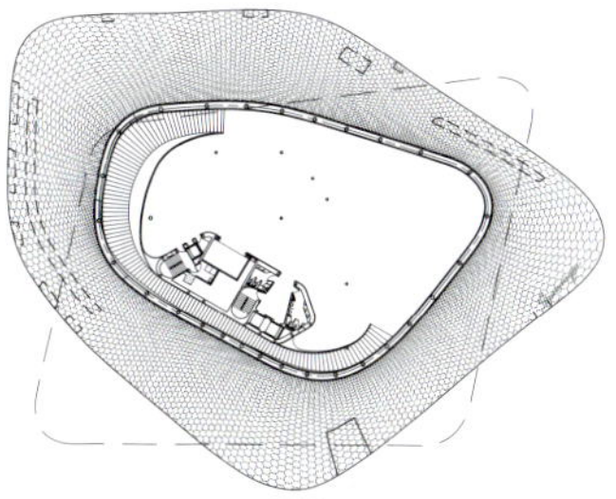

五层
fifth floor

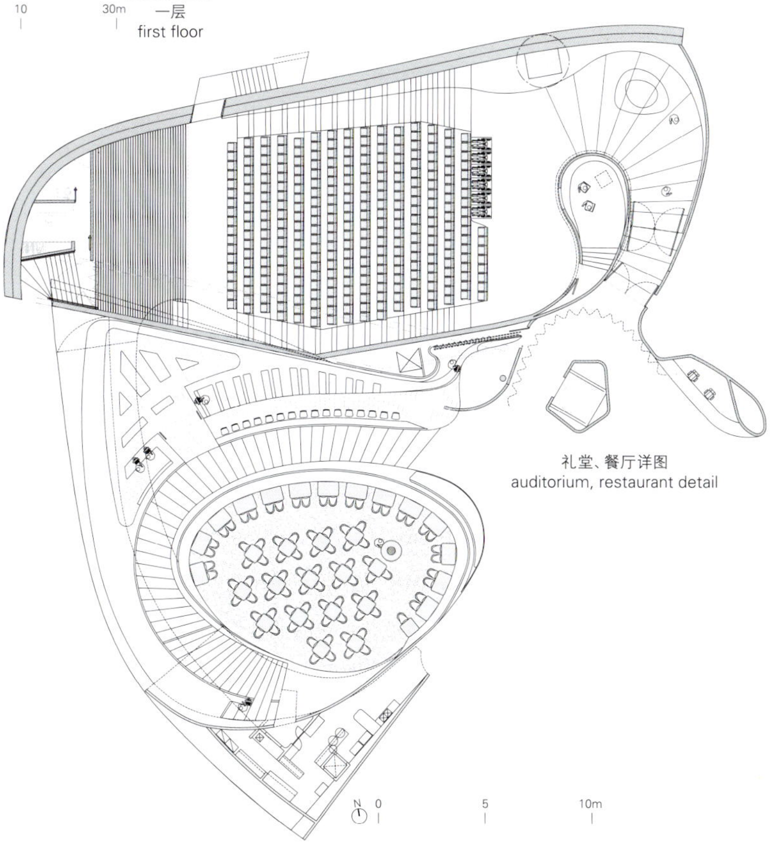

礼堂、餐厅详图
auditorium, restaurant detail

commercial activities in the neighborhood.
In order to create a new identity for the site, the proposal needed to create both a strong urban and iconic presence. Thus, the Soumaya Museum was conceived as a rotating sculptural building that is both an object and a part of the city. The museum's avant-garde shape defines a new paradigm in the timeline of Mexican and international architecture.
The shell of the building is made of 28 unique curved steel columns of varying sizes and shapes. The structure is stabilized by a system of seven rings located on each floor.
To create a shelter for its collection, the opaque facade is made of hexagonal aluminum modules that minimize exterior openings while optimizing the preservation and durability of the entire building. The conglomeration of the individual modules reflects the diversity of the museums' collection that includes the world's second largest collection of Rodin sculptures, a wing of Medieval and Renaissance Art, and a gallery of Impressionism.
This diverse collection is housed in a continuous 6,000m² exhibition space spreading over six levels. The building also includes an array of public and private programs, including: a 350 seat auditorium, library, restaurant, gift-shop, a multi-purpose lounge, and administrative offices. Connecting the programs is a non-linear circulation zone, creating areas of interaction for users along a common path as they meander through the building, culminating in a naturally lit multi-use gallery on the top floor.

FREE-Fernando Romero

项目名称：Soumaya Museum
建筑师：FREE-Fernando Romero
承建商：PC Constructores
立面：Gehry Technologies
甲方：Museo Soumaya "Fundación Carlos Slim"
地点：Lago Zurich #245, Col. Ampliacion Granada, Del. Miguel Hidalgo, Mexico
总建筑面积：17,000m²
竣工时间：2011
摄影师：courtesy of the architect-p.92, p.96, p.102 bottom, p.106
Courtesy of the architect (©Adam Wiseman)-p.99, p.103
©Sergio Pirrone-p.97, p.100~101, p.102 top, p.104, p.105, p.107

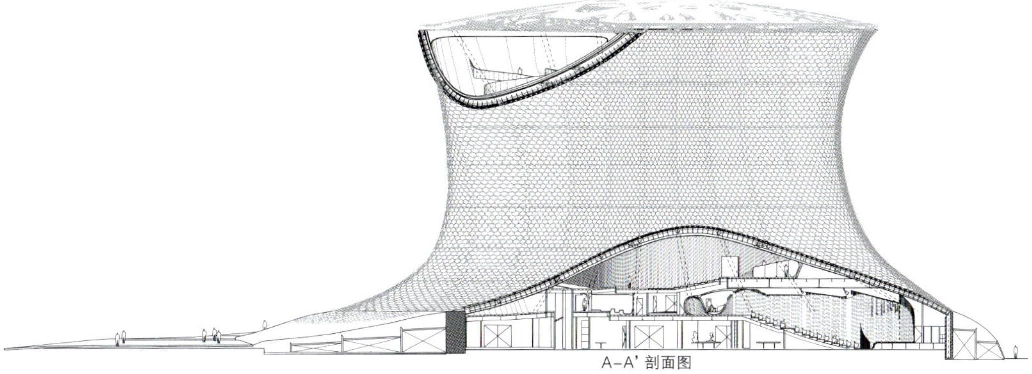

A–A' 剖面图
section A-A'

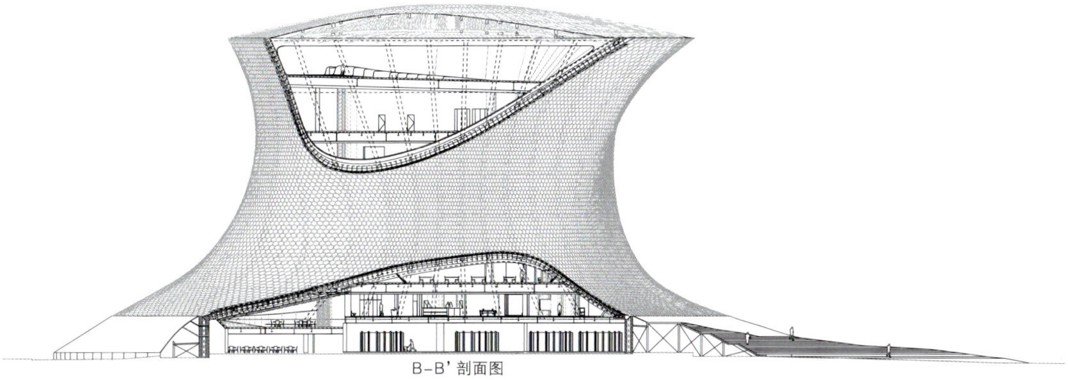

B–B' 剖面图
section B-B'

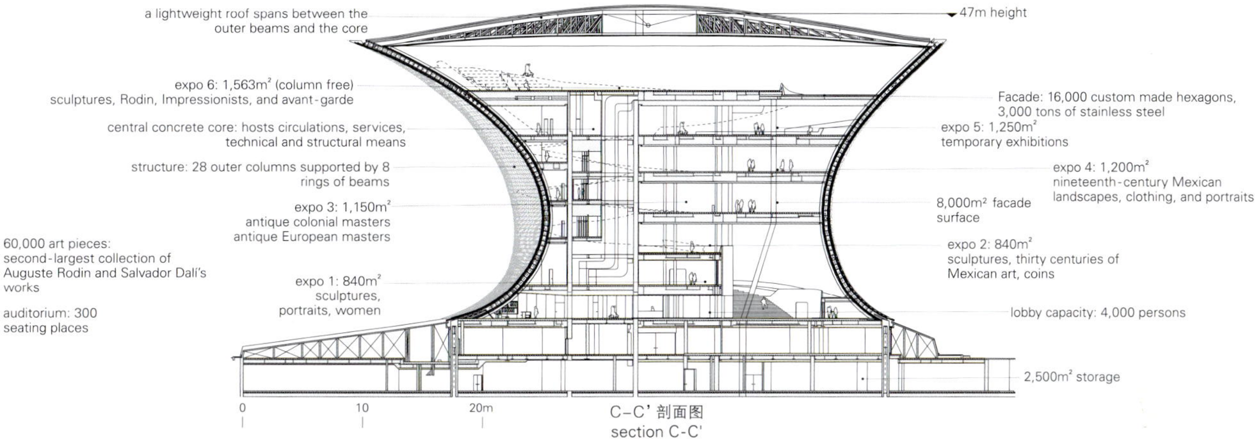

C–C' 剖面图
section C-C'

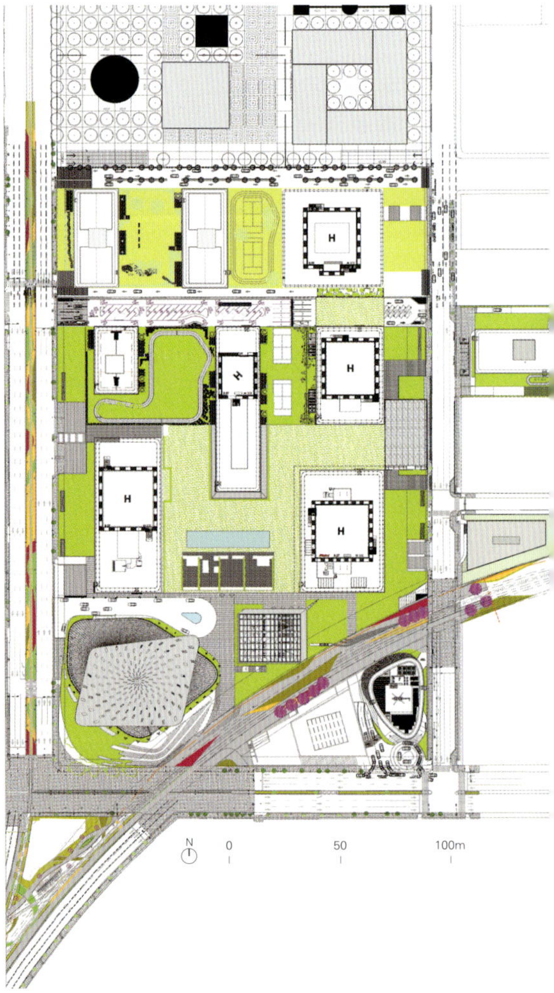
N
0
50
100m

55 20 17 71
GM
Sanborns
Banamex

安达卢西亚历史博物馆

Alberto Campo Baeza

我们想让安达卢西亚历史博物馆成为格拉纳达“最美的建筑”。建立博物馆的意图是传递安达卢西亚的全部历史。早在罗马时期，古希腊地理学家斯特雷波就将安达卢西亚的居民描述成“最文明的伊比利亚人，他们用诗文写成律法”。

我们设计的安达卢西亚历史博物馆项目与2001年建成的格拉纳达储蓄银行的总部连成一线。我们提议将其设计成一个裙楼，面积为60m x 120m，共三层，最高的一层和储蓄银行主楼的矮墙和立面齐平。建筑的布局以中心庭院为核心，这种椭圆形的中心庭院内有环形坡道，坡道从低到高，将大楼的三个楼层连接起来，形成一种十分有趣的空间张力。椭圆形庭院的尺寸参照了爱尔汗布拉宫中查理斯五世皇宫庭院的规模。

更奇妙的是，这样一个高高矗立着的垂直体，与格拉纳达储蓄银行主楼同高同宽，仿佛是城市的大门。它竖立在环绕格拉纳达市的高速路前，像一块屏幕标志牌，在巨大的等离子显示屏上向整个城市传递信息，如同伦敦的皮卡迪利广场和纽约的时代广场。

建筑的收尾处是一个延伸至河边的巨大水平平台，它是安达卢西亚历史博物馆的开放空间，也将成为格拉纳达市新区的公共空间。

这个新建筑的外形虽然安静，却用它自身的建筑元素传达着已经来到的新千年的信息。

Andalucia's Museum of Memory

We would like to make "the most beautiful building" for the Museo de al Memoria de Andalucía (Andalusia's Museum of Memory) in Granada. The project wishes to transmit the entire history of Andalusia. As early as Roman times, Strabo described the inhabitants of Andalusia as "the most cultivated of the Iberians, who have laws in verse."

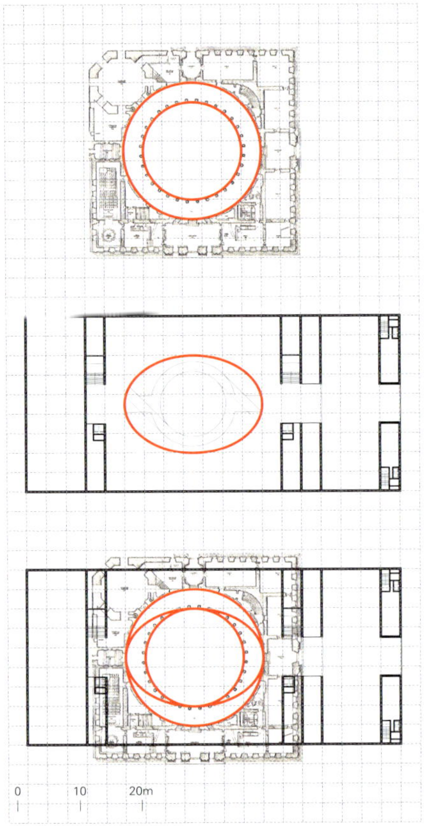

安达卢西亚历史博物馆和查理斯五世皇宫
The MA Granada and the palace of Charles V

项目名称：The MA: Andalucia's Museum of Memory
建筑师：Alberto Campo Baeza
合作者：Alejandro Cervilla García, Ignacio Aguirre López
工程师：R. Úrculo Ingenieros Consultores S.A.
甲方：Caja Granada
地点：Avenida de las Ciencias s/n, in line with the Central Headquarters of the Caja Granada Savings Bank. Granada, Spain
基地面积：15,000m²
竣工时间：2009.5
摄影师：courtesy of the architect (©Javier Callejas)

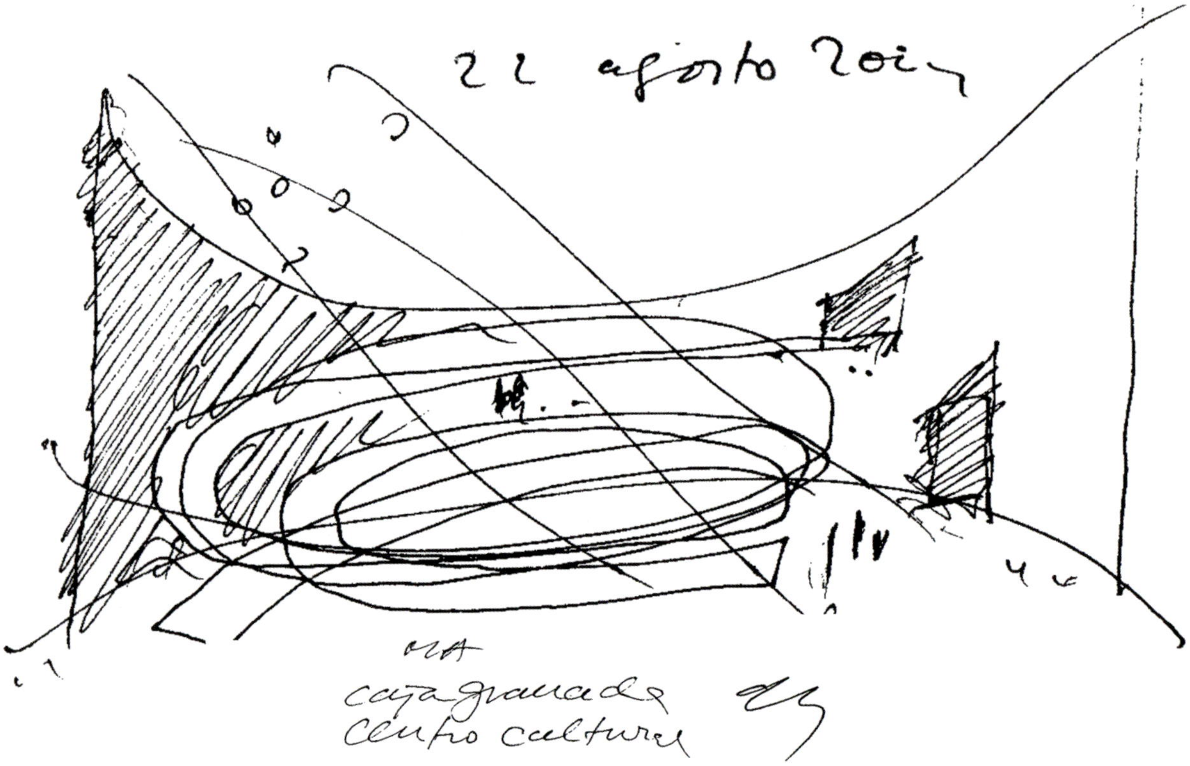

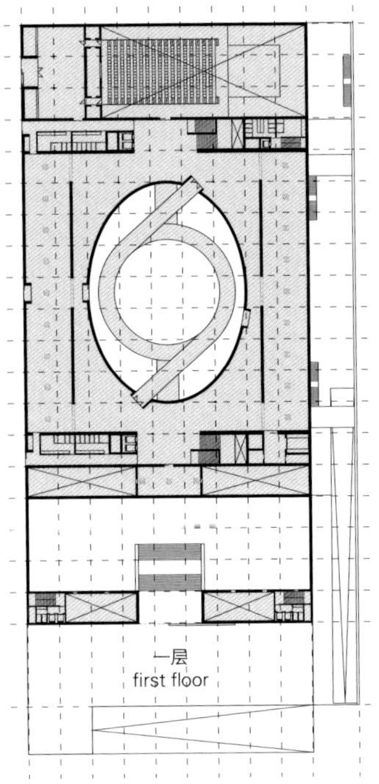

一层
first floor

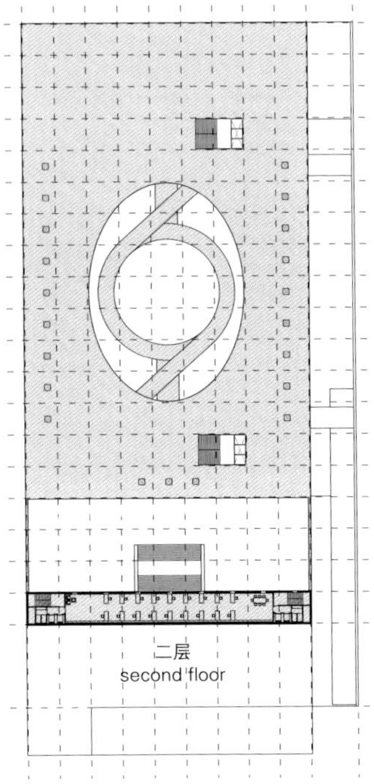

二层
second floor

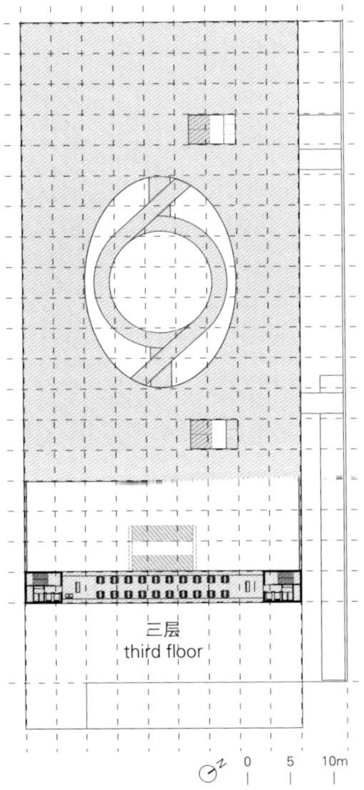

三层
third floor

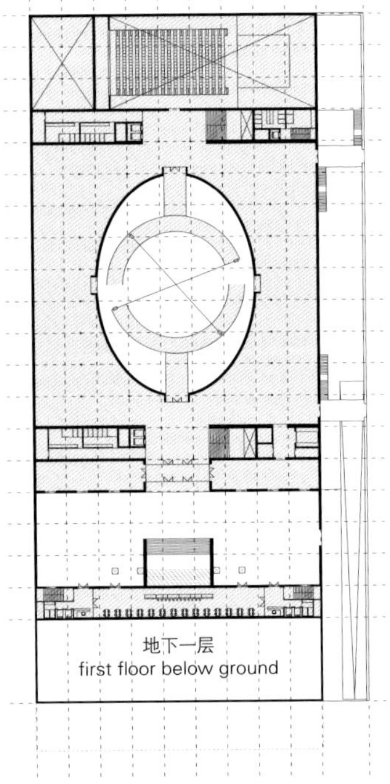

地下一层
first floor below ground

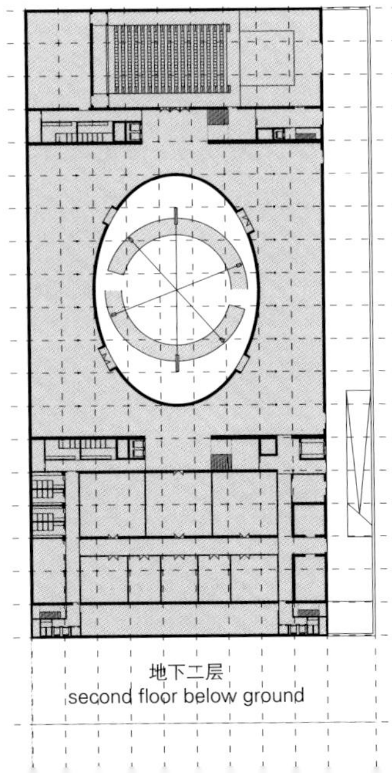

地下二层
second floor below ground

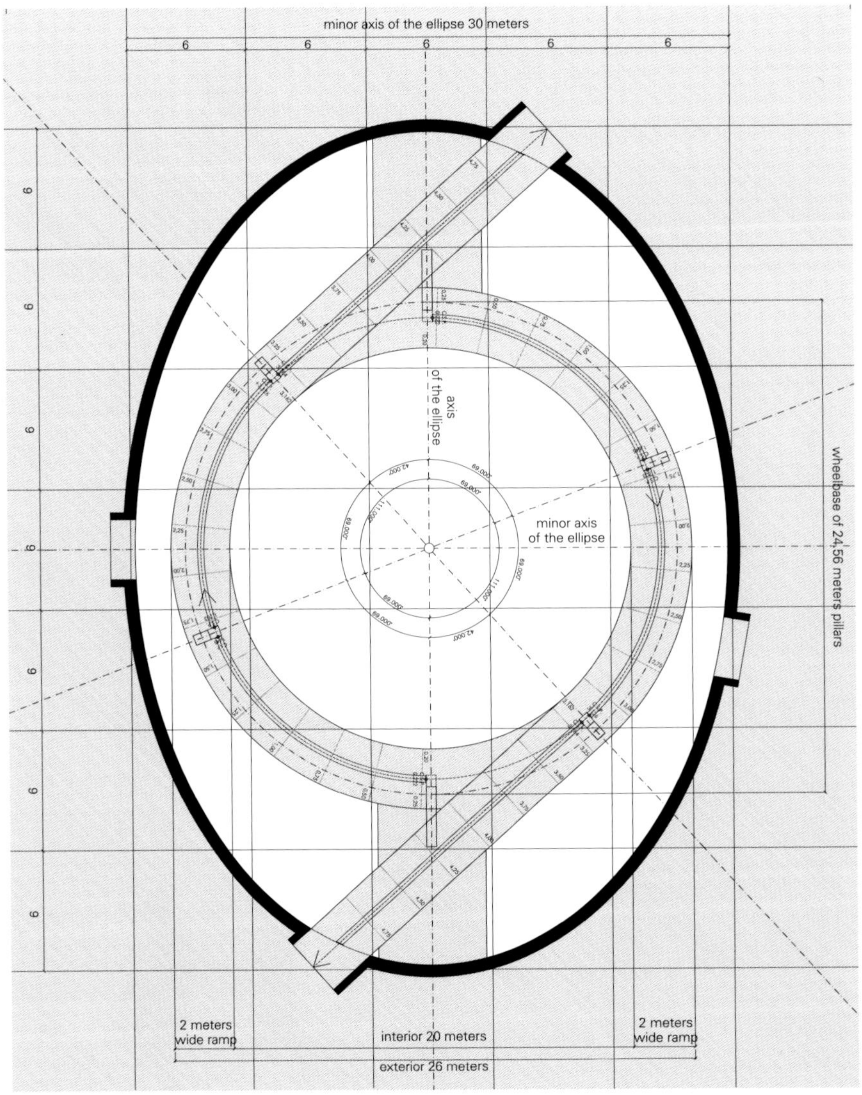

一层螺旋坡道
first floor spiral ramp

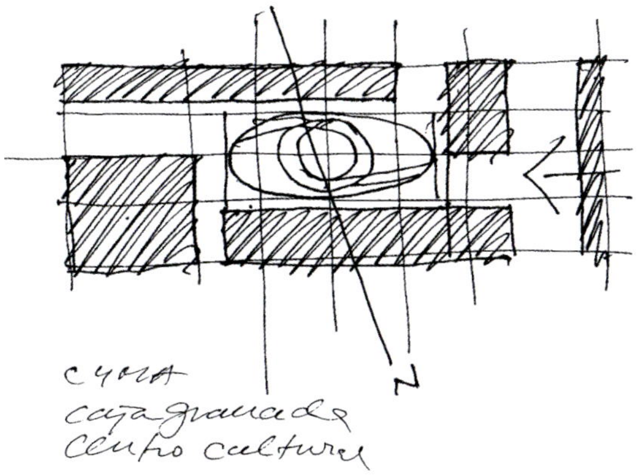
cajagranada
Centro cultural

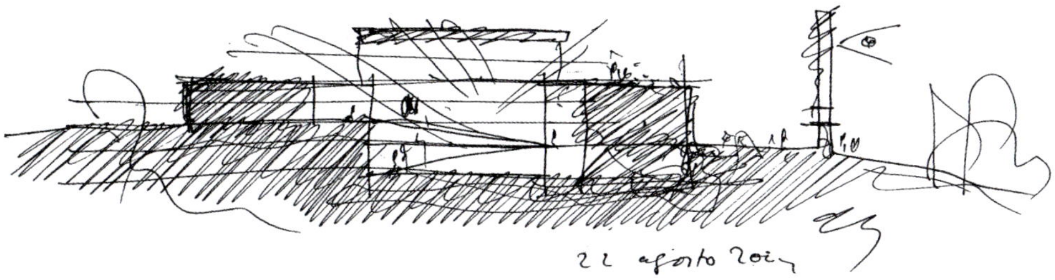

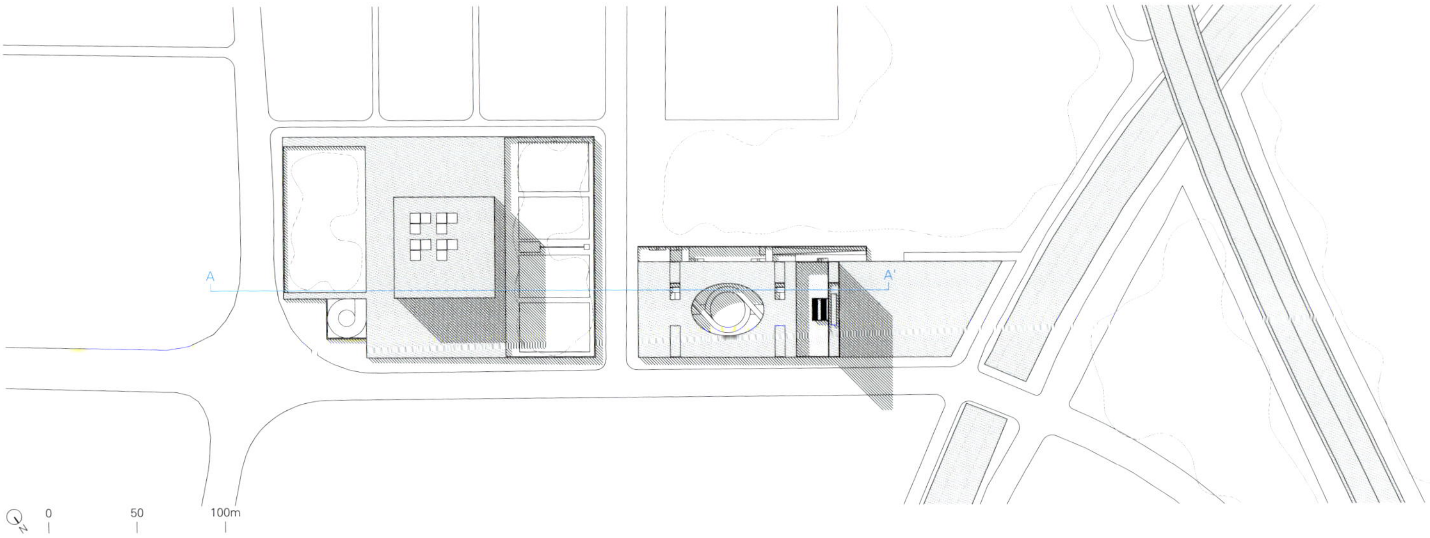

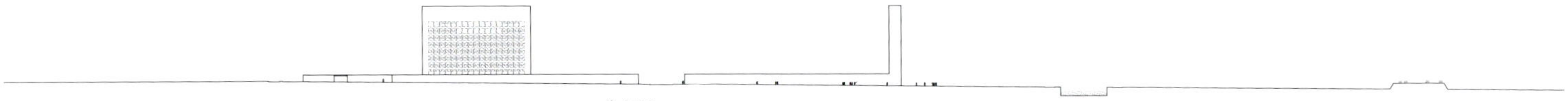

北立面
north elevation

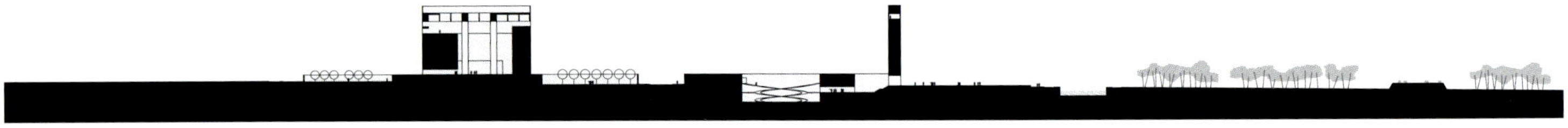

A–A' 剖面图
section A-A'

CAJAGRANADA

Our project for the MA is a building in line with the Central Headquarters of the Caja Granada Savings Bank that we finished in 2001. We propose a podium building measuring 60m x 120m and rising three stories, so that its upper floor coincides with the podium of the main Caja Granada building and its facade as well. Everything is arranged around a central courtyard, in elliptical form in which circular ramps rise, connecting the three levels and creating a very interesting spatial tension. The dimensions of the elliptical courtyard have been taken from the courtyard of the Palace of Charles V in the Alhambra.

And to crown it all, as if it were a Gate to the City, a strong vertical piece emerges, the same height and width as the main building of the Caja Granada. It thus appears before the highway that circles Granada as a screen-facade that sends messages over the large plasma screens that will cover it entirely like Piccadilly Circus in London or Times Square in New York.

And to finish the entire operation, a large horizontal platform all the way to the river is the MA open "field" that will serve as a public space in that new area of the city of Granada.

The new building, silent in its forms, is resounding in its elements to communicate the messages of the new millennium in which we are already immersed. Alberto Campo Baeza

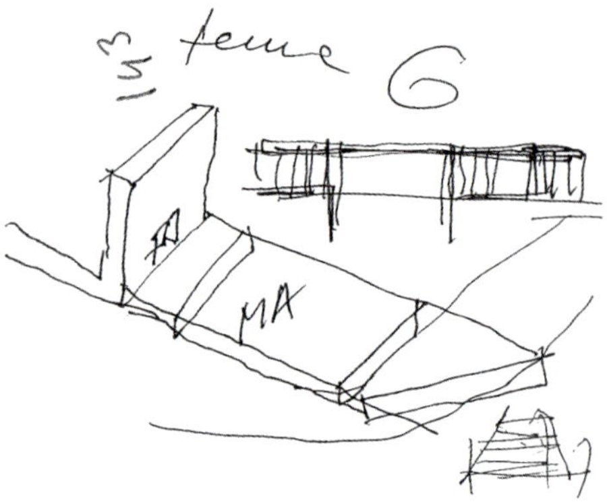

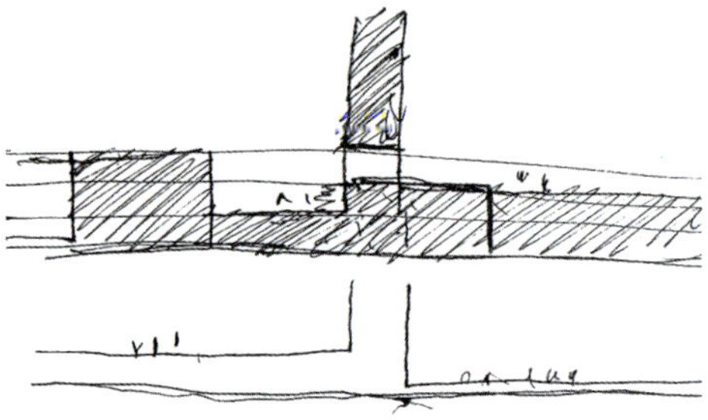

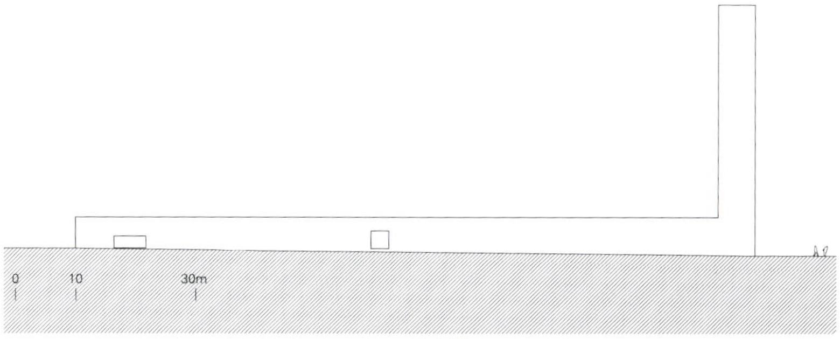

北立面
north elevation

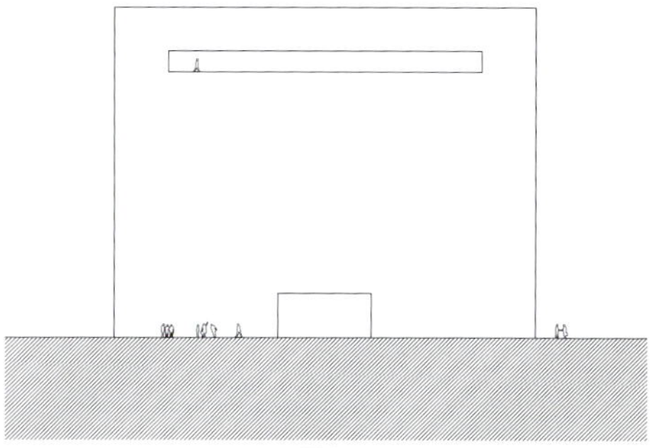

西立面
west elevation

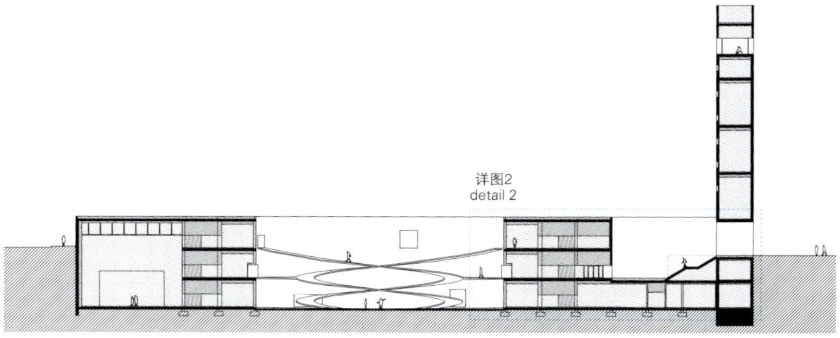

B–B' 剖面图
section B-B'

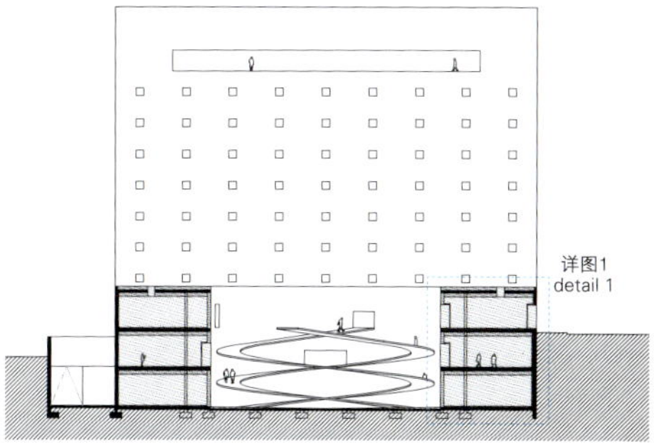

C–C' 剖面图
section C-C'

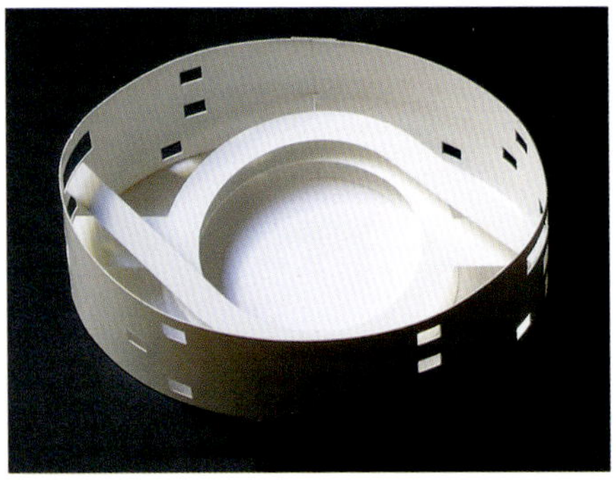

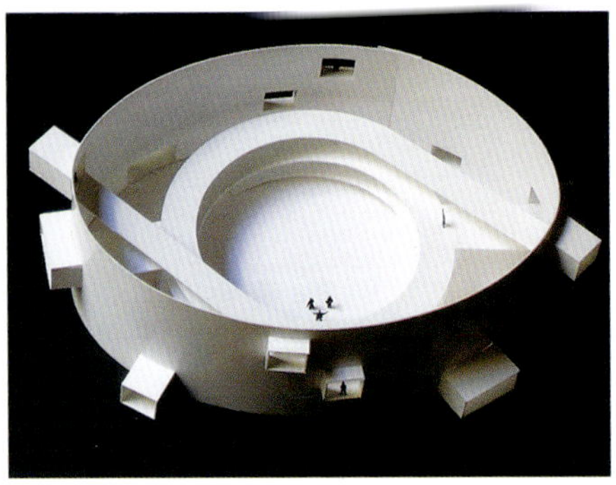

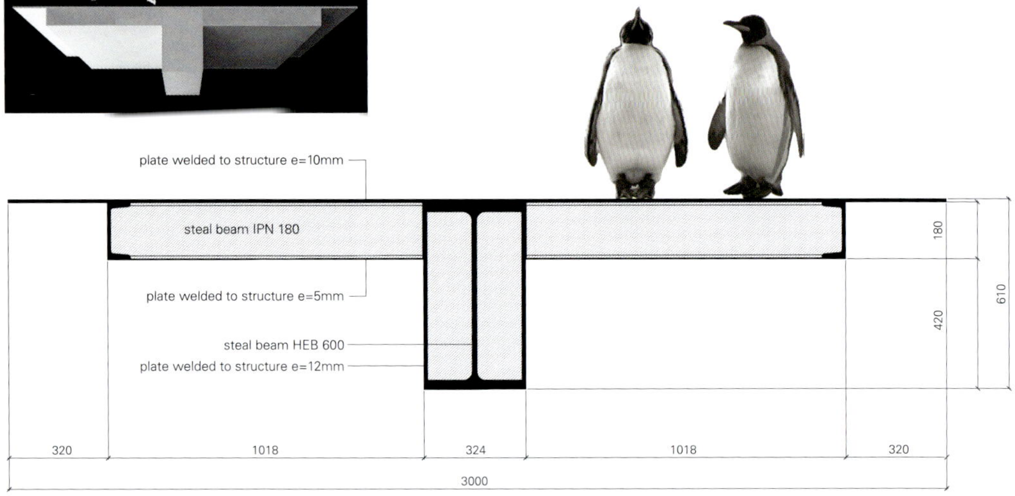

螺旋坡道剖面图
helicoidal ramp section

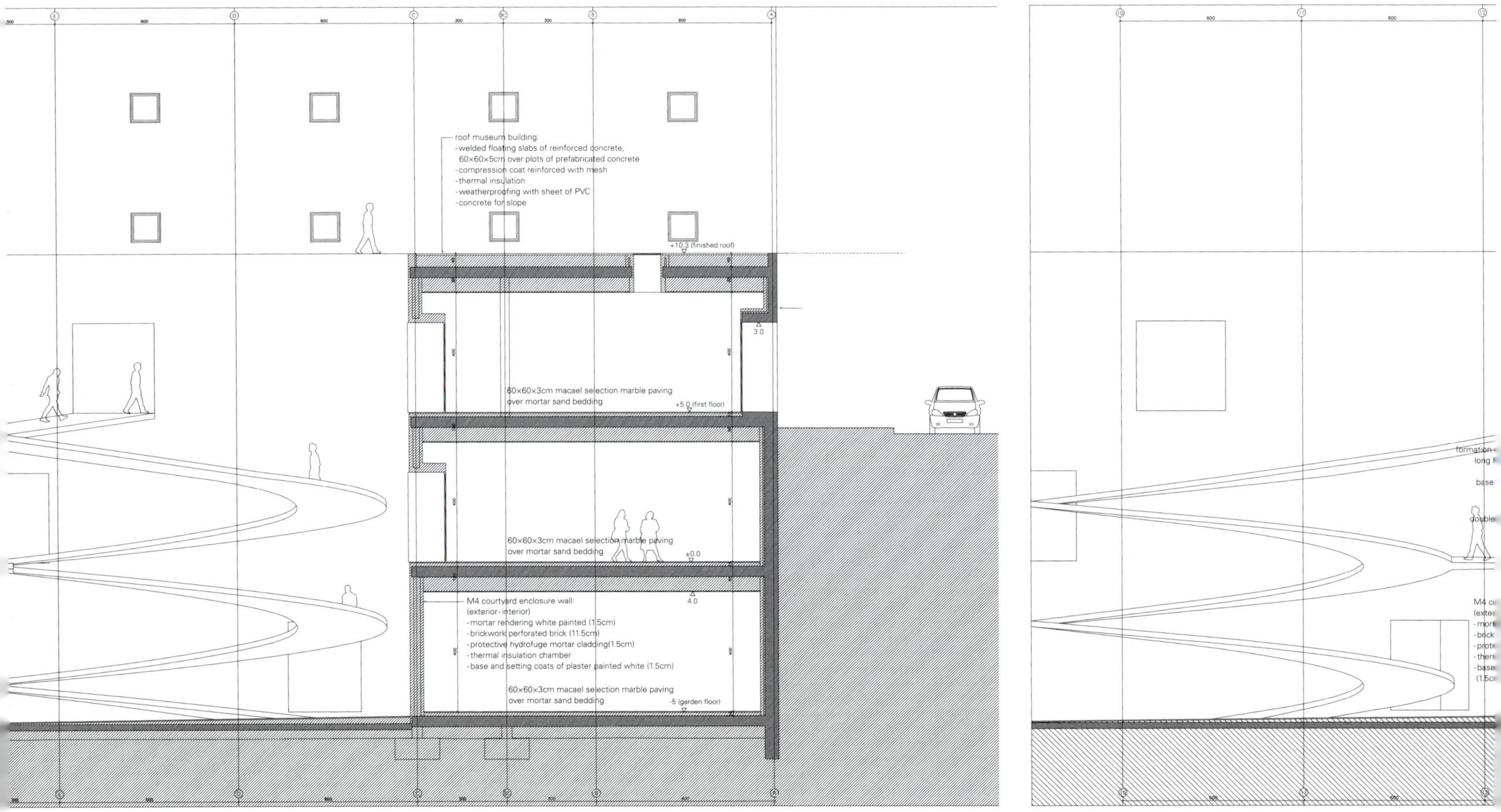

详图1
detail 1

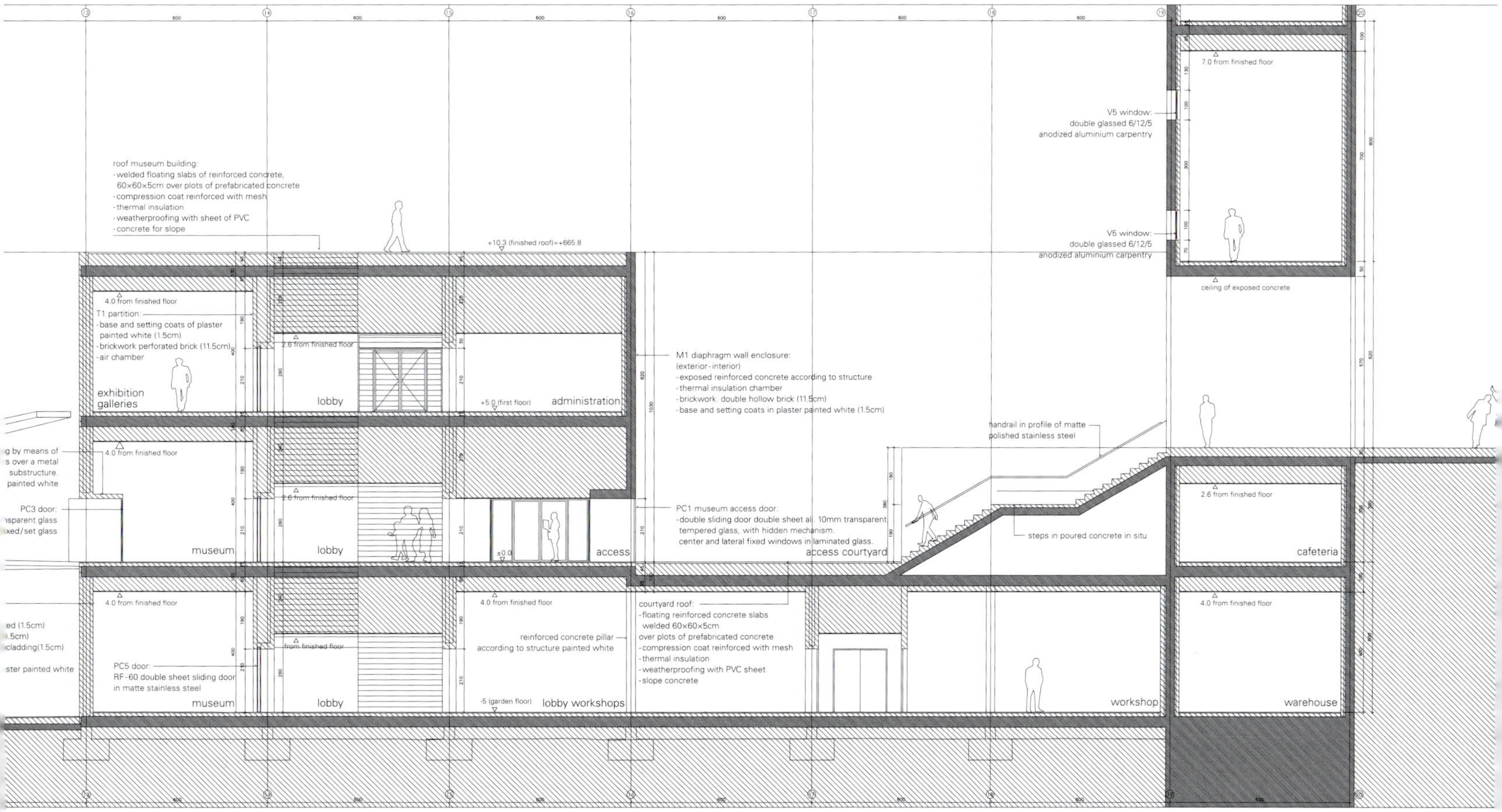
roof museum building:
-welded floating slabs of reinforced concrete, 60×60×5cm over plots of prefabricated concrete
-compression coat reinforced with mesh
-thermal insulation
-weatherproofing with sheet of PVC
-concrete for slope
+10.3 (finished roof)=+665.8
4.0 from finished floor
T1 partition:
-base and setting coats of plaster painted white (1.5cm)
-brickwork perforated brick (11.5cm)
-air chamber
exhibition galleries
2.6 from finished floor
lobby
+5.0 (first floor)
administration
M1 diaphragm wall enclosure:
(exterior-interior)
-exposed reinforced concrete according to structure
-thermal insulation chamber
-brickwork. double hollow brick (11.5cm)
-base and setting coats in plaster painted white (1.5cm)
V5 window:
double glassed 6/12/5
anodized aluminium carpentry
7.0 from finished floor
ceiling of exposed concrete
handrail in profile of matte polished stainless steel
PC3 door:
museum
±0.0
access
PC1 museum access door:
-double sliding door double sheet all. 10mm transparent tempered glass, with hidden mechanism.
center and lateral fixed windows in laminated glass.
access courtyard
steps in poured concrete in situ
cafeteria
reinforced concrete pillar according to structure painted white
courtyard roof:
-floating reinforced concrete slabs welded 60×60×5cm
over plots of prefabricated concrete
-compression coat reinforced with mesh
-thermal insulation
-weatherproofing with PVC sheet
-slope concrete
PC5 door:
RF-60 double sheet sliding door in matte stainless steel
-5 (garden floor)
lobby workshops
workshop
warehouse

详图2
detail 2

都灵汽车博物馆

改造及扩建

Cino Zucchi Architetti

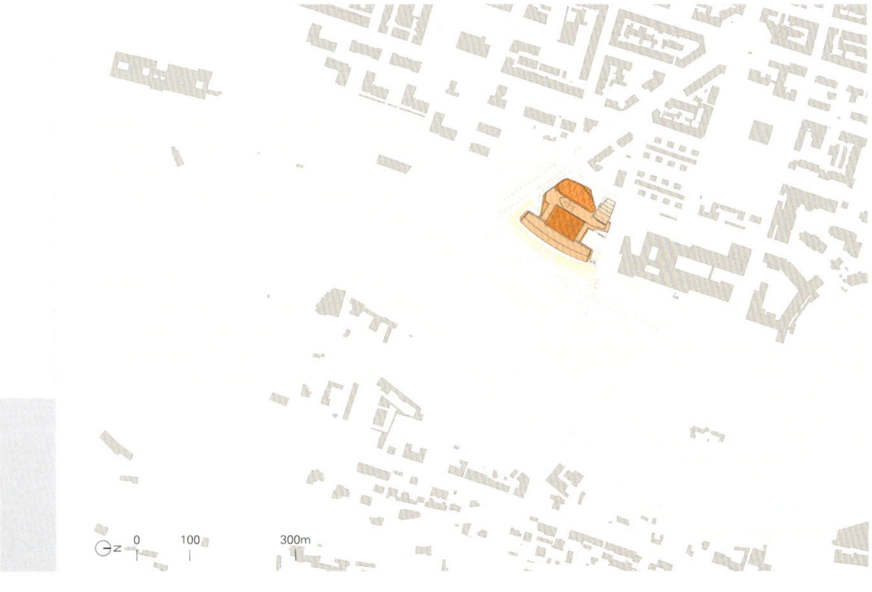

目前的博物馆是由设计师阿米地奥•阿尔贝蒂尼在1958—1960年设计建造的，它由多个部分组成，主体部分高达114m的立面面向波河和Corso Unità d' Italia。竞赛要求在Via Richelmy加建翼楼并重新调整人行道和车道，要与欧洲博物馆的最新发展相一致。新的博物馆预计每年有25万人参观，里面还将举行重要会议并开展教育活动，因此该建筑将成为该市东南部区域城市改造中的强劲驱动力。

都灵南部入口的特色在于其独特的景观及沿着波河河岸铺设的绿色林荫道，而由Pierluigi Nervi设计的拉芙宫和翻新过的帕拉维拉体育馆这两座令人印象深刻的建筑则使其形象更加鲜明。这个项目展现了取材自Corso Unità d' Italia 的“途经”理念和在通往Via Richelmy的途中建立更加亲切的步行环境的关系。全新的基座重新诠释了现有建筑的对称性，以此呼应周围的环境并强调建筑与Largo Unità d' Italia的关系。新的底层突出了河岸现有的水平线，设置了一些与之相关的空间，为公共博物馆和相关活动（书店、礼品店、酒吧餐厅）提供不同的区域。与许多欧洲同期的博物馆一样，它的展览功能融合了一系列与之互补的活动，时时刻刻为汽车博物馆注入活力。

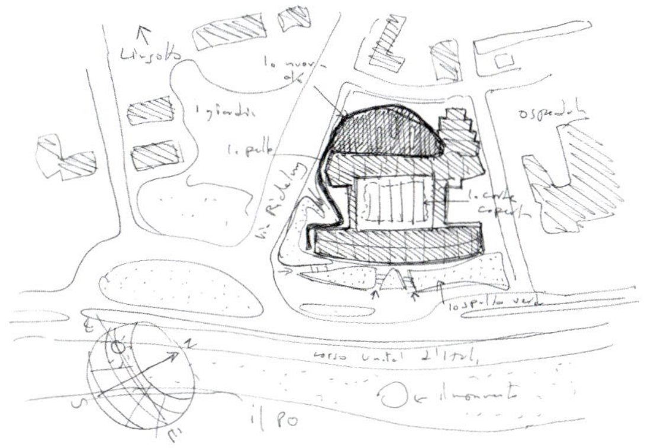

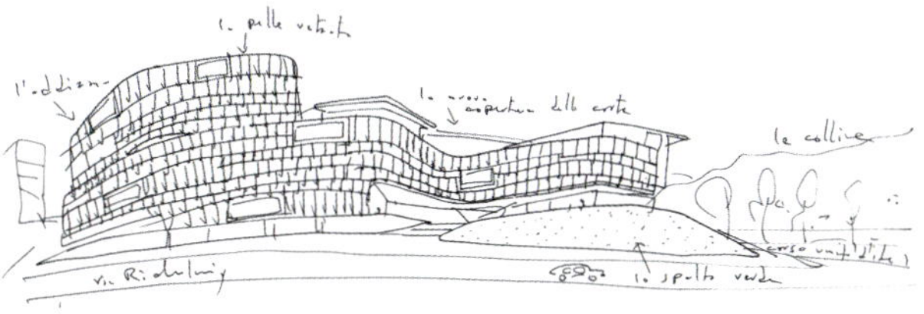

现存的庭院由于新加了玻璃屋顶，空间变得明亮起来，从而变成了一个新的活动空间。通过把现存的庭院打造成一个博物馆流线正上方的内部空间，给参观者提供了一个十分天然的定位指示。它形如“戒指”，不仅为两座主体建筑提供了连接的桥梁，还是这项设计的核心，是一项行之有效的展览方案。

博物馆西侧新的翼楼是一个完整的巨大空间，提供了非常灵活的展示空间，它和原有的建筑相融合，围绕在建筑的一侧，使两座城市建筑具有连贯性。

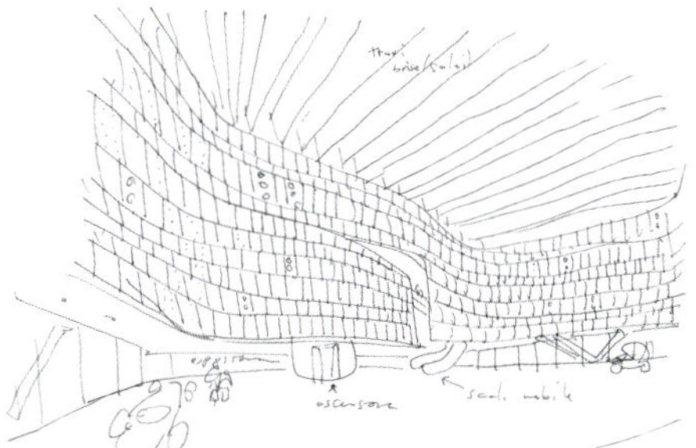

不同透明度的玻璃立面将Via Richelmy与原有建筑的立面统一起来，通过突出原有建筑和新建建筑之间的关系，赋予博物馆以新的形象，新旧建筑是通过能带来并加强建筑整体性的设计方法联系在一起的。Via Zuretti的礼堂和教室既可以单独使用，也可以和博物馆以及餐厅联合使用，从而给这座拥有欧洲最独特汽车展品的博物馆带来了新的活力。

Turin Automobile Museum
Renovation and Extension

The current museum complex, designed by Amedeo Albertini in 1958-60, is comprised of various bodies, the main section of which is a convex 114 meter facade facing onto the Po river and Corso Unità d'Italia. The competition brief stipulated the addition of a new wing on Via Richelmy and the reorganization of pedestrian and vehicular access, keeping in line with the latest development of museums in Europe. The new museum, with an expected annual attendance of 250,000 people, accompanied by significant conference and educational activities, will thus become a driving force in the urban renewal of the southeast quadrant of the city. The entrance to Turin from the south is characterized by the exceptional quality of the landscape and environment of the green parkway that runs along the banks of the river Po, punctuated

by the presence of the impressive structures of the Palazzo del Lavoro by Pierluigi Nervi and the renovated Palavela. The project articulates the relationship between the "drive-by" perception from Corso Unità d'Italia and the establishment of a more intimate pedestrian environment grafting its way on Via Richelmy. The existing building's symmetry is "redefined" by its new base, in order to respond to its surroundings and emphasize its relationship with Largo Unità d'Italia. Underlining the existing horizontal lines on the river front, the new ground floor has a number of relational spaces that accommodate the different areas of the public museum and related activities (bookshop, gift-shop, bar-cafeteria). In line with many contemporary European examples the exhibition functions are integrated with a series of complementary activities that bring life to the car museum at all hours of the day and night. The existing courtyard becomes a new event space thanks to the addition of a glass roof which illuminates the space. By making the existing courtyard an internal space, onto which the museum routes face, the visitors are provided with a very natural means of orientation. This provides not only a bridge linking the two bodies but a core at the heart of the project, taking the form of a "Ring", a proven exhibition scheme.

The new wing on the west side, a large undivided space which provides very flexible exhibition space, integrates the existing body embracing the side of the building and giving continuity to the two "urban" elevations.

The new facade of glass treated with varying degrees of transparency unifies the front on Via Richelmy, renewing the image of the museum with a great deal of attention to the relationship between the existing and new sections, connected by a design approach that can enhance and bring unity to the complex. The auditorium and educational spaces on Via Zuretti can easily work both independently and in conjunction with the museum and the restaurant, animating this museum complex hosting one of the most unique collections of cars in Europe. Cino Zucchi Architetti

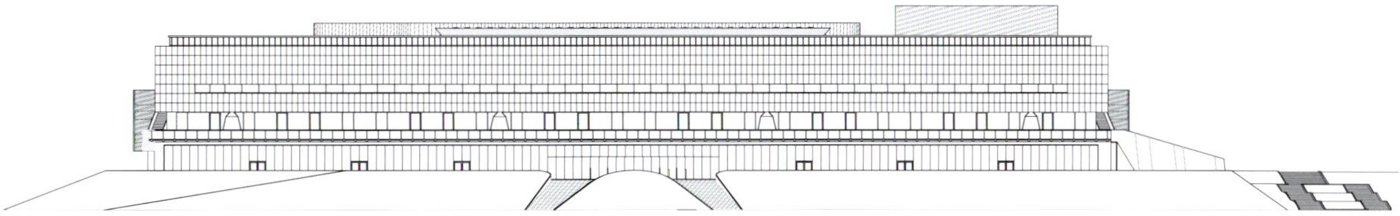

东南立面
south-east elevation

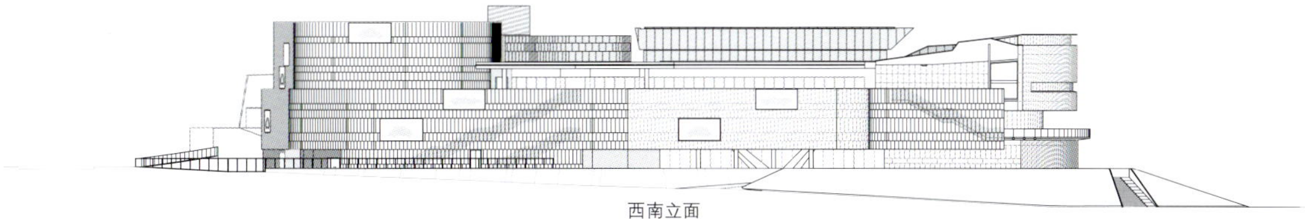

西南立面
south-west elevation

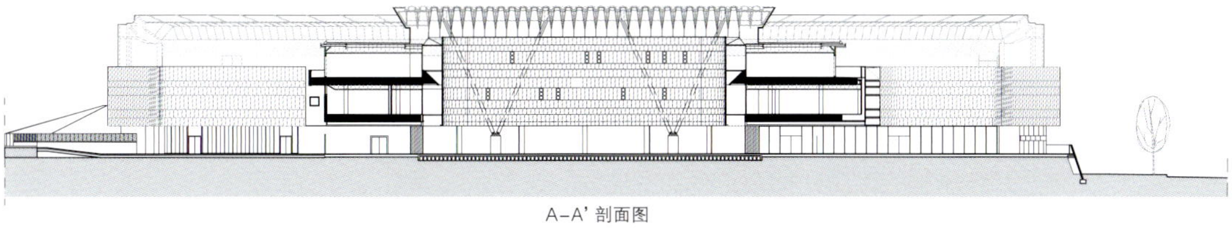

A-A' 剖面图
section A-A'

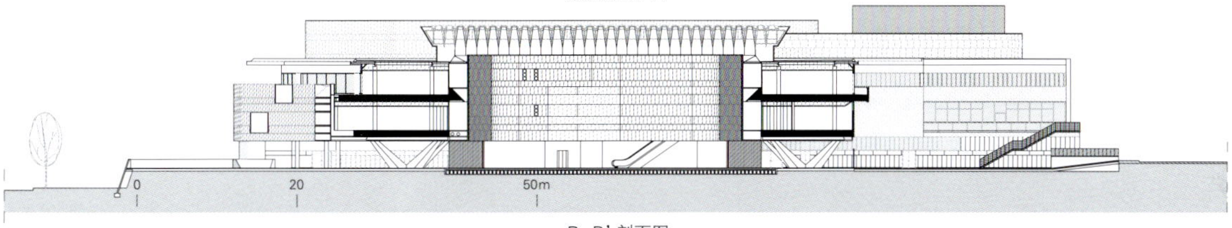

B-B' 剖面图
section B-B'

项目名称：National Automobile Museum in Turin renovation and extension
建筑师：Cino Zucchi Architetti
工程师：Proger Spa
修复工程顾问：Michela Catalano
展览设计：François Confino and Studio LLTT
地点：Turin, Italy
基地面积：17,700m²
施工时间：2005~2011
摄影师：©Filippo Poli

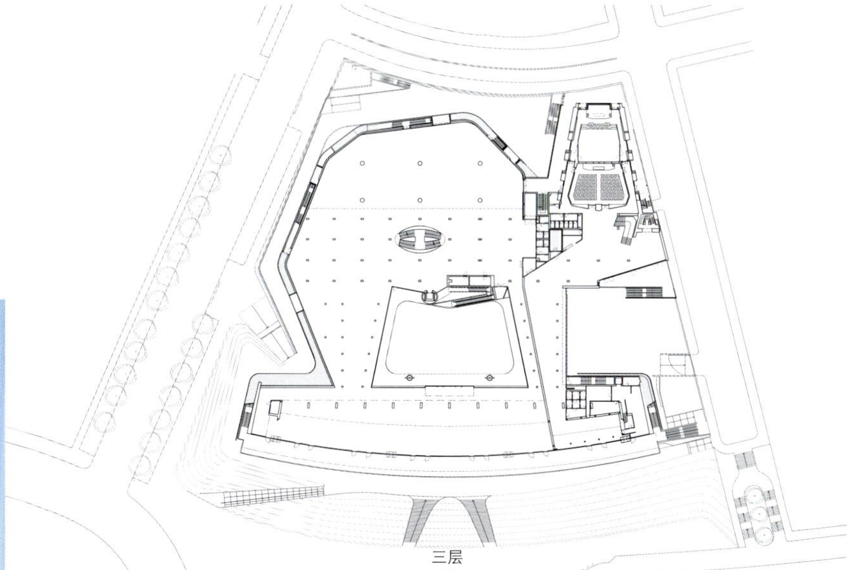

三层
third floor

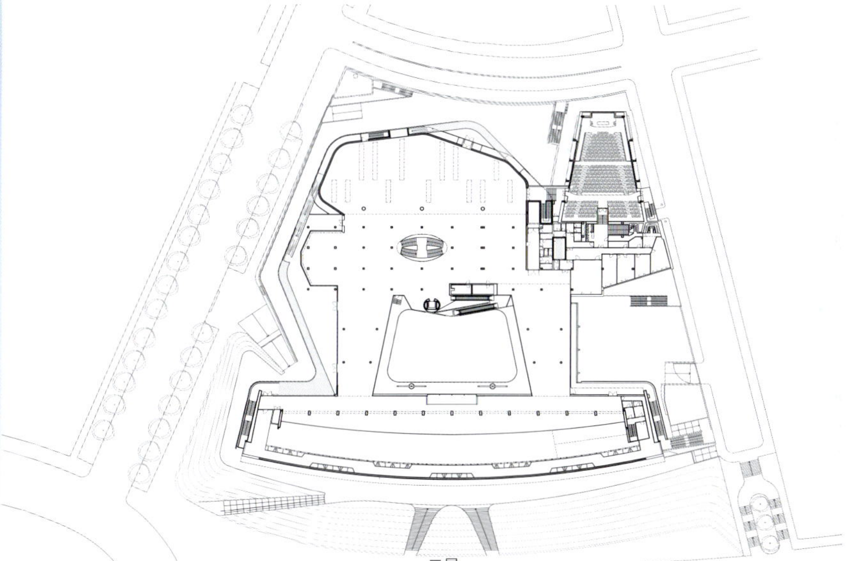

二层
second floor

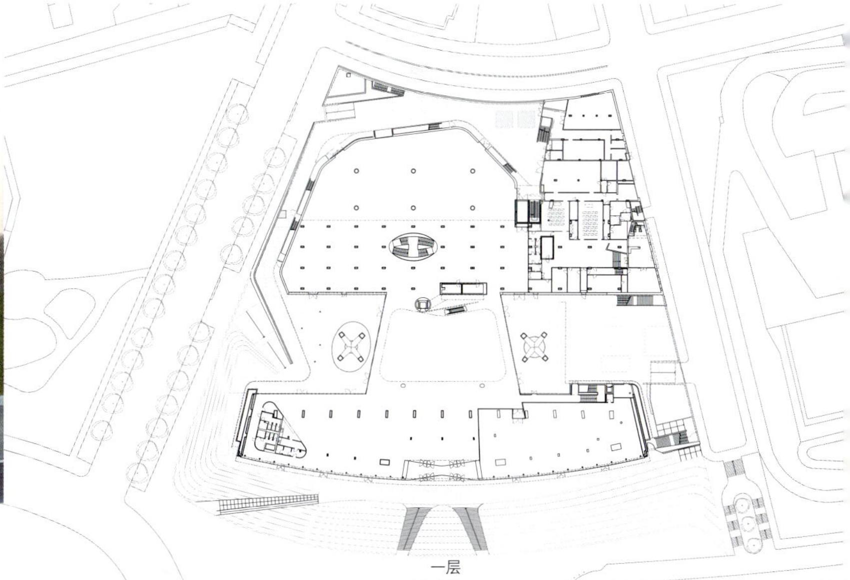

一层
first floor

0 20 50m

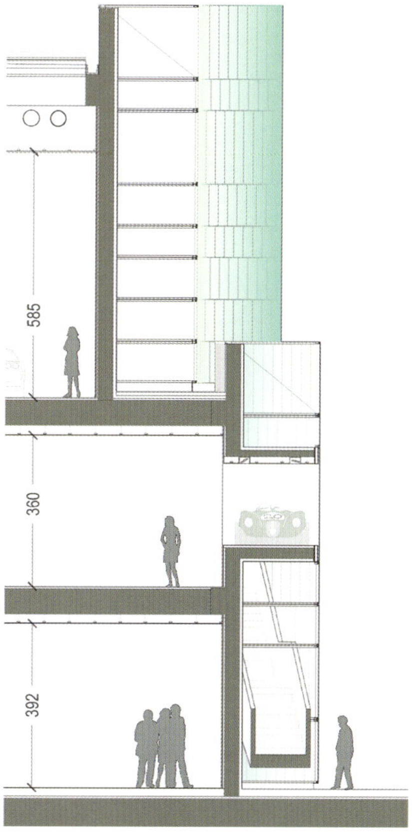
585
360
392

20

表演是种高度协作的艺术形式。它是艺术家和观众之间的互动和对话。针对每个事件的思想交流是独特的。传统的礼堂为观众和表演者提供了空间，但是文化的交流不应该仅限于舞台表演。教室、门廊和大厅都能提供类似的交流机会。而且在这些自由的空间里，艺术家和观众不会被传统的剧院布局隔离开来。在外面的广场和大厅，任何人都可以成为表演者，任何人都能体验到表演的乐趣。文化中心提供参与的平台，努力使文化体验成为人们日常生活的一部分。文化中心通过这样的方式改变着它们所在的社区。

As an art form, performance is highly collaborative. It is a social interaction and dialogue between the artist and viewer, where the exchange of ideas is unique to each event. Traditional auditoriums provide a space for the performers as well as the audience, but cultural exchange isn't necessarily limited to a traditional performance stage. Classrooms, hallways and lobbies provide a similar opportunity for communication. Moreover, it is in these unconventional spaces that the artist and the audience are not separated by the usual layout of the theater. In outside plazas and lobbies everyone can be the performer, and everyone can experience the work. As such, cultural centers can transform their community by offering a forum for participation, and striving to make cultural experiences a part of everyday life.

艺术中心的新概念：作为参与和感知的平台
New Conception of Art Center as a Forum for Participation and Awareness

聚在一起

随着人口密度的增加，我们变得比以前任何时候都更加个性化。我们就像孤独的岛屿，虽然住在一起但却不常联系，也不常以团体的形式聚在一起。但是，我们一直都在分享我们的文化活动。文化活动没有歧视，不会质疑我们的信仰或背景，而是欢迎我们每一个人融入到公共的世界中来。它们就像一张交错的网，面对着这张网，我们全体投降，坠入其中。这一刻，我们能暂时放下自己的喜怒哀乐，走到一起去体验外部的场景。

表演作为一种艺术形式，对合作的要求很高。它是艺术家和观众之间的互动和对话。针对每个事件的思想交流是独特的。传统的礼堂为观众和表演者提供了空间，但是文化的交流不应该仅限于舞台表演。教室、门廊和大厅都能提供类似的交流机会。而且在这些自由的空间里，艺术家和观众不会被传统的剧院布局隔离开来。来到外面的广场和大厅不需要购买门票，也没有指定的座位。任何人都可以成为表演者，任何人都能体验到表演的乐趣。像这样，剧院和艺术中心等文化中心通过提供参与和感知的平台，改变着它们所在的社区。

法国莱赛比耶的艺术之塔已经成为该区的文化标志。该区有两座历史悠久的塔，一座属于教堂，另一座代表市政厅。这两座塔俯瞰着欧洲城市。它们是日常生活的一部分。居民在这座城市的任何地方都能看到它们。它们提醒居民要尽的公民和道德义务。建造新音乐学校的同时，莱赛比耶城建造了第三座塔，它成了一个文化标志。

艺术之塔由Forma 6设计，并在2010年竣工。它位于Place des Droits de l' Homme的人权广场。它包括一所设有音乐系、舞蹈排练室和礼堂的艺术学院。高塔的西立面朝向公共广场，有六层高。与塔

Coming Together

With increasing density we have become more individualized than ever. We are solitary islands, living alongside each other and only rarely connect or come together as a community. However, what we continue to share are our cultural activities. These events aren't discriminatory; they don't question our beliefs or background, but welcome us into a public world. They act as a connecting net, in which we can collectively surrender and let ourselves fall. For a short while, we're not concerned with our own lives, and can come together to experience an exterior scene.

As an art form, performance is highly collaborative. It is a social interaction and dialogue between the artist and viewer, where the exchange of ideas is unique to each event. Auditoriums provide a space for the performers as well as the audience, but cultural exchange isn't limited to a traditional performance stage. Classrooms, hallways and lobbies provide a similar opportunity for communication. Moreover, it is in these unconventional spaces that the artist and the audience are not separated by the usual layout of the theater. Accessing outside plazas and lobbies doesn't require a ticket, and here, there is no assigned seating. Everyone can be the performer, and everyone can experience the work. As such, cultural centers, such as theaters and art centers can transform their community, offering a forum for participation and awareness.

The Tower of Arts in Les Herbiers, France has become a cultural icon for its community. Traditionally two main towers, one belonging to the church and the other one marking the town hall, dominate European cities. These towers have a presence in daily life. They are visible from any part of the city and remind its inhabitants of their civic and moral duty. With the new music school, the city of Les Herbiers created for a third tower, a cultural symbol, this time.

The Tower of Arts, designed by Forma 6 and completed in 2010, sits on the Place des Droits de l'Homme, a square dedicated to human rights. It houses an arts academy with a music department, dance rehearsal spaces, and an auditorium. A tall tower rises six stories on its western facade, which faces the public square. Juxtaposed with the broad angled volume of the tower is the build-

Forma 6提供
courtesy of Forma 6

艺术之塔，主入口
Tower of Arts, main entrance

楼巨大的倾斜体量相呼应的是建筑的入口。入口是一个长方形的混凝土立方体。倾斜的赤陶红色天花板衬托着玻璃门。参观者由玻璃门进入。进入门内是座位区，供在这宽敞的两层高大厅内打发时间的参观者使用。

排练厅设在建筑的南半部分，教室在北侧。每间教室都备有特定的乐器或演奏组，进而营造出更加明显的特征。每种音乐都有自己的故乡，无论是非洲的打击乐还是“世界音乐”，它们相互配合成为声音的团体。

教室之间的空间形成了社交环形区。建筑师用少量的抛光漆来凸显这个负面空间。入口处就清晰可见的裸露混凝土就像常春藤一样在建筑里蔓延，延伸到所有的楼层和每一个角落。在结构上和装饰上引进树状的主题使得这个空间的自然延伸名副其实。中庭的内墙被刺破来营造混凝土“树”，树枝向天花板蔓延，并与中庭通往二楼的楼梯隔离开来。排练工作室设计成三个白色的箱体，像树屋一样挂在树结构的格子上，让空间充满嬉闹的童趣。空间采用了最小化的进一步装饰。建筑师设计了律动的穿孔图案来装饰大厅里白色的排练包厢。红色材质的使用突出了学校礼堂的入口。这种明亮的颜色在室外得到延续，塔的立面上宽宽的红色切口意外地打破了铜骨架外墙。金属立面和突兀的红色点缀的结合意在提醒参观者，这幢建筑不仅是一个教育机构，还提供更多的东西，它在努力营造文化氛围浓郁的社区。

西班牙维克的阿特兰蒂达表演艺术中心是现有的音乐学校的新扩建部分。原有建筑“肯•塞拉托萨”现在包括新校的公共区域，而新建筑增加了剧院、礼堂和餐厅，将这个区域改造成一个更大的充满艺术气息的校园。在这幢建筑里，建筑师Josep Llinàs将各种规划好的

ing's entrance, which is defined by a rectangular concrete cube. A sloping terracotta-red ceiling highlights glass doors through which visitors enter. Inside the doors lies a seating area for anyone who comes in and wants to spend some time in the large double height lobby.

The rehearsal rooms are located in the south half of the building, and the classrooms are housed in the north. Each of the rooms is assigned to a specific musical instrument or group in order to create a stronger sense of identity. Each type of music has its home, whether it is African percussion or "World Music", but they work together as a community of sounds.

The spaces between the educational rooms form a circuit of social exchange. The architects accentuated this negative space with spare finishes. The bare concrete that was first seen in the entry cube extends through the building like ivy, reaching to all floors and into every corner. The natural spread of this space is made literal by introducing an arborescent theme, both in structure as well as decoration. The interior walls of the central atrium are punctured to create concrete "trees", that branch up to the ceiling and provide a partial separation from stairs that lead up to the second floor. Three white boxes, rehearsal studios, hang like tree houses from the structural latticework and give the space a childlike playfulness. Further ornamentation is minimal. The architects designed a rhythmic pattern of perforations that decorate the white rehearsal boxes in the lobby and the use of primary-red fabric emphasizes the entrance to the school's auditorium. This bright color is continued on the exterior, where a broad gash in the facade of the tower creates an unexpected break in the copper cladding. The combination of the metallic facade with the theatrical red accents reminds visitors that this building is an educational institution first, but that it also has more to offer and strives to create a rich cultural community.

The L'Atlantida Performing Arts Center in Vic, Spain, is a new extension to an existing music school. The older building, the "Can Serratosa," now houses the common areas of the new school, while new construction adds a theater, an auditorium and a restaurant, transforming this area into a larger artistic campus. With the building, architect Josep Llinàs has merged various programmatic volumes under one faceted zinc roof. The complex is located between the proper boundaries of Vic on the north and an

Josep Llinàs提供
courtesy of Josep Llinàs

阿特兰蒂达表演艺术中心北入口
north entrance of L'Atlantida Performing Arts Center

部分都纳入一个片状锌皮屋顶下。建筑的北面是维克固有的边界，南面是未来建筑的新区。Llinàs用建筑的形状为这两个邻近的区域搭建了一座桥梁。沿着维克的外围，项目的一部分被埋入地下，新屋顶遵守城市的高度限制。米堤亚河将老城与新的扩展区隔离开来，屋顶越靠近米堤亚河越高。这样当它被新的城市建筑所包围时，新的立面便会成为文化中心的前沿。

起伏的屋顶平面形成了意想不到的庭院，使自然光能够照射到音乐学校和餐厅，但礼堂等不需要自然光的空间则被安排在室内区域。通过使用铜铝合金和TECU金铜提高了庭院空间的亲切感。金色营造出奢华的内墙空间，这与该工程的黑色混凝土外墙截然不同。金铜一直延续到室内，勾勒出剧院空间墙壁的轮廓。正如艺术之塔所表现的，金色和红色搭配的观众席使人们感受到传统剧院的氛围。但是，通过在外面延续这样的颜色搭配，Llinàs证明了用于公众集会的空间也可以用来进行文化交流，人们参观这里可以得到的不仅仅是传统的体验。

艺术之塔和阿特兰蒂达表演艺术中心丰富了城市的学术和艺术文化。它们体现了艺术和休闲的教育价值。像艺术中心一样，剧院可以极大地丰富地区的文化生活。Mrpr Arquitectos设计的Infanta Leonor剧院位于西班牙南部，距离马德里有四个小时的路程。剧院是沿着哈恩省的小镇的古城墙建造的。作为伊比利亚半岛最古老的城市之一，小镇保存了重要的考古遗址。城中到处都是罗马时代的城市住宅区遗迹。中世纪时，哈恩还是摩尔人的格拉纳达和欧洲其他的基督教王国之间的边界。如今，它可以作为再现这些多层面文明的景观，模

area of future construction to the south. Llinàs has used the building's shape to create a bridge between the two neighborhoods. Along the outside perimeter of Vic, part of the program has been buried in the ground and the new roof adheres to the height limit of the city. As it moves towards the Meder River that separates the old city from the area of future expansion, the roof level increases to allow for a new facade that will become a front for cultural center when new urban construction surrounds it.

The undulating roof plane shapes unexpected courtyards that allow for natural light to reach the music school and restaurant, while spaces that don't require natural light, like the auditorium rooms, are fit into the interior areas. The intimacy of the courtyard spaces is heightened by the use of the copper-aluminum alloy, TECU Gold. Its gold color creates rich inner spaces that are dramatically different from the exterior walls of the project, which are finished in black concrete. The gold is continued on the interior, where it lines the walls of the theater spaces. As is seen in the Tower of Arts, the use of gold with the red color of the audience seating makes for a traditional theater experience. However, by continuing this color combination outside, Llinàs has proven that communal gathering spaces can also lend themselves to cultural interaction, and visiting this complex will give more than a traditional experience.

The Tower of Arts and the L'Atlantida Performing Arts Center have added to the academic and artistic culture of their cities. They represent the educational value of the arts and leisure. Just like art centers, theaters can greatly enrich a community's culture. The Infanta Leonor Theater by Mrpr Arquitectos is located in the south of Spain, four hours south of Madrid, where the theater sits along the old city walls of a small town called Jaén. As one of the oldest cities of the Iberian Peninsula, the town preserves an important archeological heritage. Remnants of urban settlements dating back to Roman times remain throughout the city and in the Middle Ages, Jaén was the border between Moorish Granada and the Christian kingdoms of the rest of Europe. Today, it can be interpreted as a landscape of these multiple layers of development, blurring the distinction between its rich history and the present.

The Infanta Leonor Theater design was selected as the winner in a competition that called for a theater that would integrate remains of the historic Puerta del Aceituno, one of the ten original gates

Infanta Leonor剧院西侧俯视图
downward view of Infanta Leonor Theater from the west

糊了悠久的历史和现代之间的区别。Infanta Leonor剧院设计在一场竞赛中取胜。这场竞赛是要选出能将Puerta del Aceituno——古城墙十个古城门之一——的历史遗留整合起来的剧院。Mrpr Arquitectos在他们的提案中阐述了关于历史整合和优化的两个主要意图。

首先，他们提供参观者免费参观的机会。其次，他们着手将古城门的遗址转变成象征性的入口，大众可以在那儿体验和学习该区的历史和文化。

Mrpr设计的建筑由三个截然不同的部分组成，包括主大厅和两个礼堂。停车场、办公室、排练厅和化妆间等服务区域在下一层。Puerta del Aceituno的遗迹被纳入了建筑结构中，由带隔板的门廊保护，公众可以从建筑的北面街道进入。

从外观上看，三个规划好的结构迥然不同。在这座古城的城中心区，建筑有着棱角分明的设计，与该地不规则的形状相呼应。长长的两层高大厅位于两个礼堂的中间区域，并一直向外面的广场延伸。公共露台环绕在建筑的四周，使得古城墙和剧院新的石灰岩外墙形成不间断的对比。参观者被街道引领着，经过沿着一个陡峭的斜坡而下的高高的石墙，逐渐映入眼帘的便是剧院了。这些流线正是建筑外观鲜明轮廓的写照。此外，流线在建筑的室内也清晰可见，在室内，剧院包厢的承重墙向建筑的中心稍稍倾斜。室内这些微小的变化使得空间体验，尤其是从室外到室内的体验充满了活力和情感。剧院吸引着街道上的参观者走进这个公共空间，来到空间内部观看演出，来了解这座城市的历史。通过这些设计，建筑师们成功地实现了他们为自己设定的两个目标。

距离哈恩省西部几小时的路程还有另外一家新的剧院，剧院于

in the old city wall. In their proposal Mrpr Arquitectos stated two main intentions for historical integration and enhancement.

First, they requested free access to all visitors. And second, they set out to transform the remains of the portal into a metaphorical gateway for the community to experience and learn about the region's history and culture.

Mrpr's building is made up of three distinct volumes, housing the main lobby and two auditoriums. Service spaces such as parking, offices, rehearsal spaces and dressing rooms are located on a lower level. The remnants of the Puerta del Aceituno were incorporated into the architecture and are covered by a screened porch that is publically accessible from the street that marks the north border of the site.

The forms of the three programmatic boxes are left distinct on the exterior. Their angular plans are carefully arranged to fill the irregular shape of this site in the heart of the historic city. The elongated double height lobby sits as the central form between the two auditoriums and extends out towards the exterior plaza. This public terrace surrounds the building on all sides and gives a continuous contrast between the old city walls and the new limestone clad exterior of the theater. Visitors are guided in from the street as they pass high stone walls that decline along a steep slope to meet the sidewalk, gradually revealing a view of the theater. These lines mirror the sharp forms in the building's exterior. In addition, they are visible on the inside of the building, where the containment walls of the theater boxes lean towards the building's center at a slight angle. Carrying through the motif of these subtle shifts in to the interior makes the experience of the space, from outside to inside particularly dynamic and emotional. There is a strong attraction from the street into the public space and into the interior of the theater, inviting all to come together for the theater's performances, and to learn about the city's history. With their design, the architects have successfully achieved both of the goals they set for themselves.

A few hours west of Jaén lies another new theater, which opened in 2009. Architect Enrique Krahe consolidated old buildings in a new skin to form the Municipal Theater of Zafra. The vernacular style is recognizable in the new design, and defines a unique character for this transitional site, located between the old city and new construction on its outskirts. To smoothly merge these two

扎夫拉市政剧院　舞台连接礼堂内外
Municipal Theater of Zafra. Stage is seen connecting inside and outside of the auditorium.

z©M. Guzmán)提供
courtesy of Enrique Krahe (©M. Guzmán)

2009年开业。建筑师Enrique Krahe为原有建筑增加了新的建筑表皮，形成扎夫拉市政剧院。新设计具有浓郁的本土风格，使这个传统的地区形成了自己的特色。该区位于老城和外围新建筑之间。为了将这两个邻近的地区完美地结合起来，公众的可接近性就变得尤为重要。室外空间有铺着鹅卵石的行人专区和露台。

设计该剧院不仅是为了便于大众参观，而且给大众提供演出之外的东西。简单的设计元素，如用于指路的照明设备，能够将参观者的注意力转移到公共空间上。观众可以在礼堂观看表演，但是建筑还邀请参观者做一回演员。在意想不到的地方有很多细小的幽默元素，如卫生间玻璃上彩绘的胡子图案或座椅的颜色显露出的眼睛，这些幽默元素都让人忍俊不禁。但更重要的是，建筑让公众意识到，剧院的重要性不限于带有舞台和观众席的建筑。在这儿，舞台墙壁可以对外开放，这样演出就变成了全城的体验。此外，建筑的外墙上布满了成对出现的小型圆窗，甚至是卫生间的小便池上方也有，这样参观者可以见到外面的景色，仿佛是在提醒他们，任何人随时随地都可以将自己想象成一名演员或是更大的文化中的一员。

作为文化和艺术中心，这些工程都努力使文化体验成为我们日常生活中的一个真实的部分。它们强调公共空间的社会重要性，吸引当地人和参观者进入社区并提供重要的公共对话。建筑是表演的舞台，更是交流的平台。在那儿，每个人都可以是表演者、参与者或者观看者，这体现了建筑本身和参观者对周边地区社会生活的重要性。

neighborhoods, public accessibility was particularly important, and the outdoor spaces are highlighted as pedestrian zones with cobblestones and terraces.

The theater has been designed to be easily visited by the community, but also to provide them more than just its shows. Simple design elements, such as lighting used as way-finding, turn the attention of visitors to the communal spaces. The audience can spend time watching the performances in the auditorium, but the building also invites the visitor to be an actor. Small humorous elements in unexpected place, such as a mustache painted on the bathroom mirror or an eye revealed in the colors of the audience chairs, make for a comic experience. But more importantly, the building offers its public an unusual thought – the importance of a theater isn't limited to the building with a stage and audience seating. Here, the stage walls can open to the outside, making the shows a city-wide experience. In addition, small binocular windows pierce the building's exterior wall, even above the urinals in the bathroom, to give visitors views to the landscape outside, as if to remind them that anyone can consider him or herself as an actor and part of a larger culture, anywhere and at any time.

As cultural and artistic centers, all of these projects strive to make cultural experiences a true part of our everyday life. They highlight the social significance of public spaces that invite locals and visitors into the community and provide important communal dialogue. The buildings offer a stage for reflection, but also a platform for discussion. They are spaces where everyone can be a performer, a participant, or a viewer, making its visitors as well as the buildings themselves valuable additions to their neighborhood's social life. Julia van den Hout

Julia van den Hout 生于阿姆斯特丹，于2004年移居纽约，并就读于纽约大学。获得了艺术史和城市设计的学士学位之后，她供职于Steven Holl Architects。现任出版和展览经理，处理纽约和北京办事处的出版业务。她参与了视觉艺术学校设计批评部门的研究，重点关注建筑批评和荷兰设计。

Julia van den Hout was born in Amsterdam and moved to New York in 2004, where she attended New York University. After graduating with a Bachelors degree in Art History and Urban Design, she started work at Steven Holl Architects, where she is currently the Press and Exhibitions Manager, handling press for both the New York and Beijing office. She is involved in studies at the Design Criticism department at the School of Visual Arts, with a focus on architecture criticism and Dutch Design.

阿特兰蒂达表演艺术中心

Josep Llinàs

我们明白工程应当与地理位置的特殊性和项目的需要相结合。该项目位于目前的综合城镇和将来扩建区的边界上。米尔德河滋养的一条狭长自然地块突出了从一座建筑向另一座建筑的过渡。这个综合城镇的北边由许多庭院构成，从联排住宅的背立面可以俯视庭院。

在这些庭院之间，我们发现了“肯•塞拉托萨”——维也纳国际中心音乐学院现在的家，它将成为新音乐学校的公共区。

“肯•塞拉托萨”有一条能从街头直接进入场地的小巷。这成为设计发展中的一个重要方面。

另一方面，场地南面边缘的建筑边界没有任何真正的连续性，我们认为通过给未建场地和由河流平原划定的绿地增加连续性，可以解决这个问题。

就项目来说，它的基本组成部分是剧院、礼堂、服务厅、音乐学校和餐馆。

剧院和礼堂是大尺度且不需要自然光的单元，而音乐学校却是重复叠加的小单元并需要自然采光和通风。对于剧院而言，舞台塔不可避免的高度是一个突出特点。它的体量为城市和场地周边区域有关的建筑提供了视觉参考。

我们做草案时贯彻了以下目标：

把综合城镇中的建筑看做城市构造的一部分。然而，另一方面，将开放空间连接到由河岸规划的自然区域里。

运用可以使用相同的设计方法（基本上是与地面的关系、几何结构和屋顶）的工程系统来解决用途和尺度全然不同的设计，就像构成本项目的各个组成部分这样。与此同时，成功地把塔台放入这个应用系统支配的范围内。

从这种意义上来说，从“肯•塞拉托萨”通往外部空地的路变得非常重要，它就像一根脊柱一样，为项目的每一个部分定位并提供到达的通道。从舞台塔顶部倾斜下来的屋顶一直延伸，直到它覆盖住门廊和教室。

L'Atlantida
Performing Arts Center

We understand the project responds to the particularities of its location and program needs. The site is placed between the boundary of the consolidated town and the future expansion areas. The transition from one fabric to the other is articulated by a strip of natural land supported by the Meder River. In the northern edge, the consolidated town is composed by courtyards where the rear facades of row housing look over.

Among them, we find "Can Serratosa", the current home of the Vic Music School, which will hold the common areas of the new music school.

"Can Serratosa" has a side lane that permits direct access from the

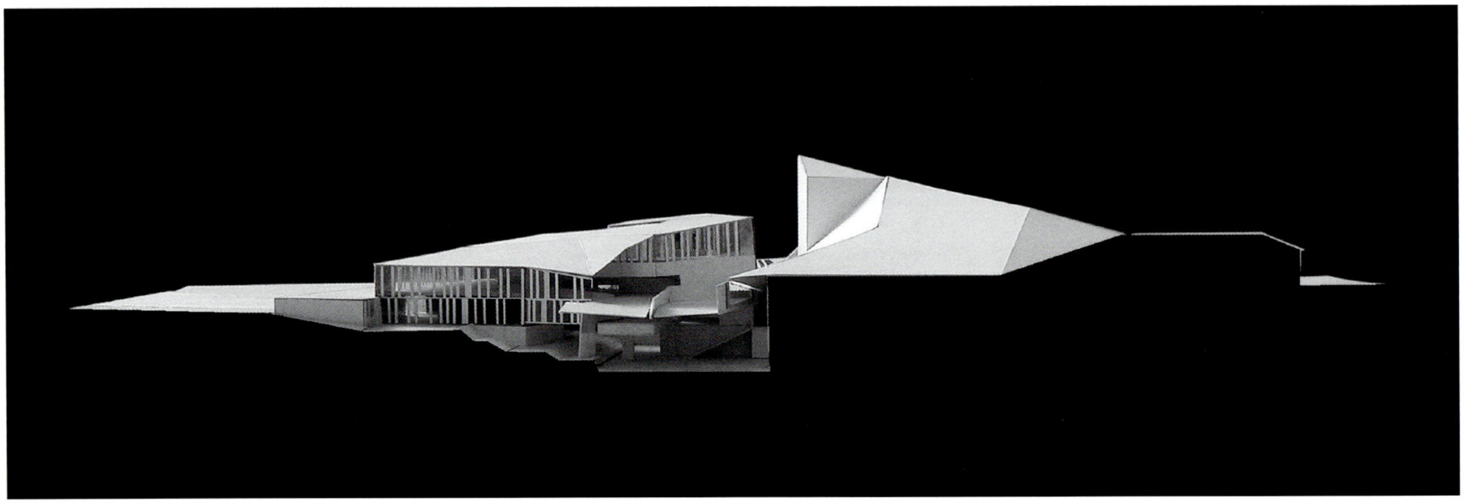

N
0
200
500m

street to the site. This becomes an important aspect in the development of the proposal.

The southern edge, on the other hand, only has a property boundary without any real consistency, which we think should be dissolved using the continuity solution for the unbuilt site and the green area defined by the river flats.

In relation to the program, it is basically made up of parts of different constitution: theater, auditorium, services, music school and restaurant.

While the theater and the auditorium are large-scale units with no need of natural light, the music school is the repetitive addition of small units that need natural light and ventilation. In the case of the theater, the inevitable height of the fly tower is an outstanding feature. Its volume makes it a visual reference for the complex in relation to the city and the local environs.

We bore the following aims in mind while drafting the proposal.

Regarding the building ascribed to the consolidated town as a melted part of the urban fabric. While, on the other hand, linking the open space to the natural territory organized by the riverbank.

Using project systems that would be with the same designing tools – basically the relationship with the ground, the geometry and the roofs – to resolve items with utterly different uses and scales like the ones that form the program. And, at the same time, managing to include the fly tower volume in the rules of the system applied.

In this sense, it has been quite important for the use of the lane from "Can Serratosa" to the open space as a backbone that situates and provides access to every part of the program, and also the use of the sloping roof which descends from the top of the fly tower and spreads out until it covers the porches and classrooms.

Josep Llinàs

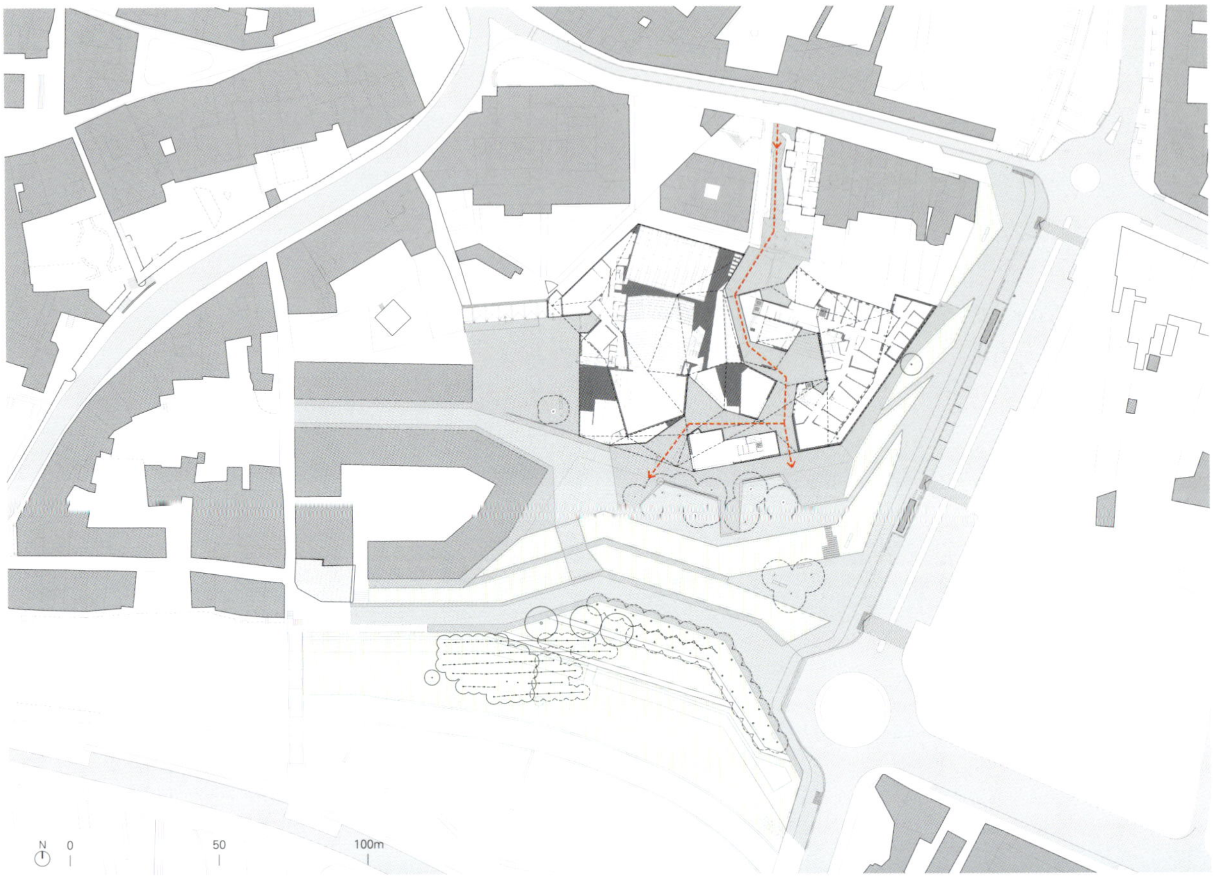
N
0
50
100m

L'Atlàntida
Centre d'Arts Escèniques d'Osona

立面1 elevation 1

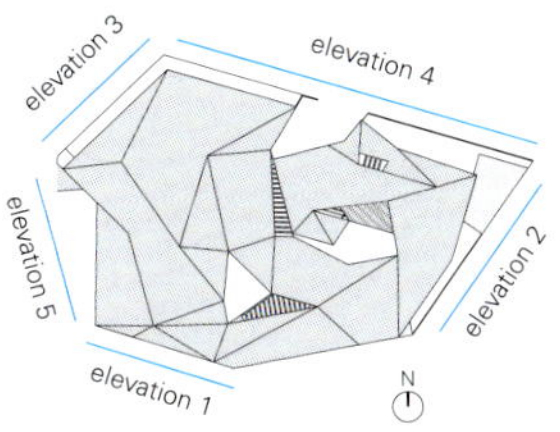

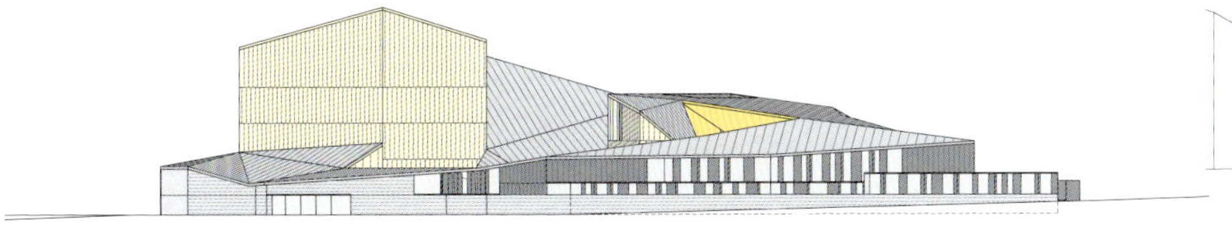
立面2 elevation 2

立面3 elevation 3

立面4 elevation 4

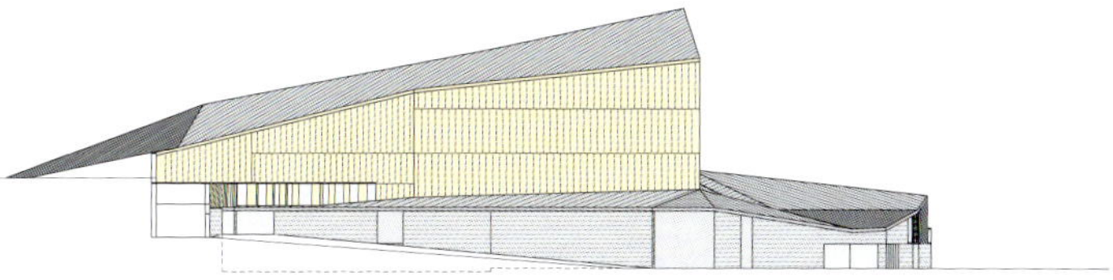
立面5 elevation 5

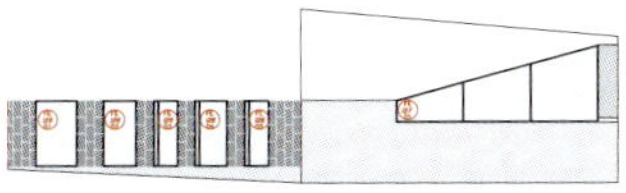
立面1展开图 unfolded facade 1

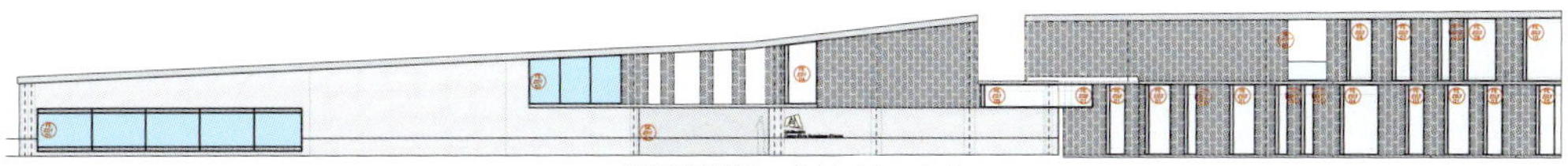
立面2展开图 unfolded facade 2

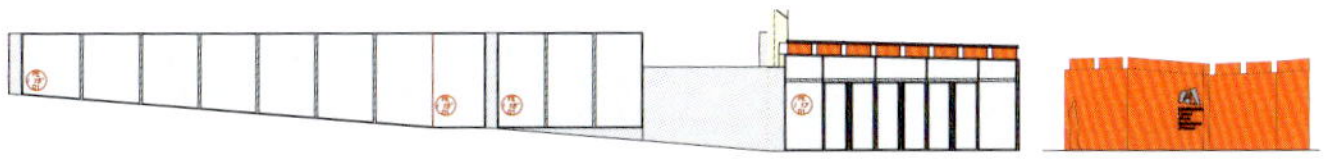
立面3展开图 unfolded facade 3

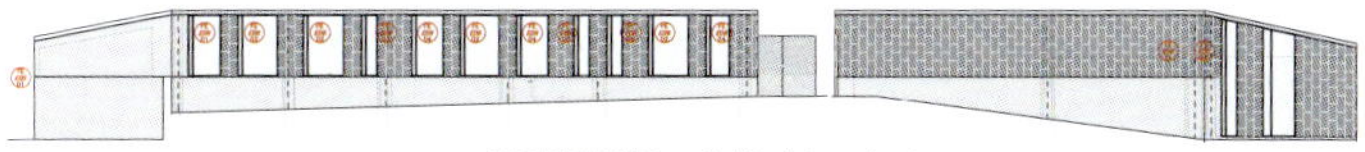
立面4展开图 unfolded facade 4

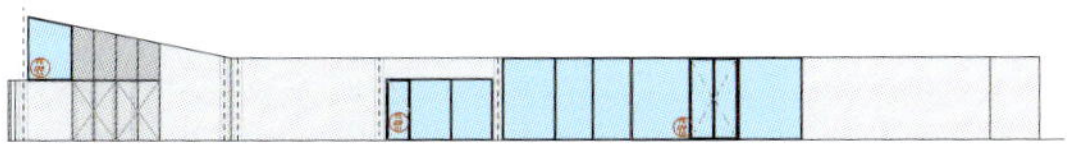
立面5展开图 unfolded facade 5

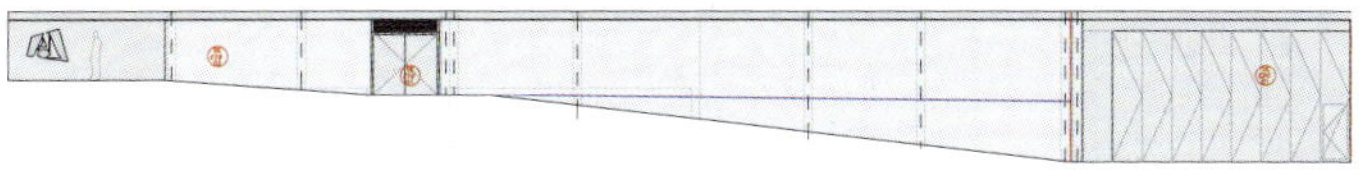
立面6展开图 unfolded facade 6

立面7展开图 unfolded facade 7

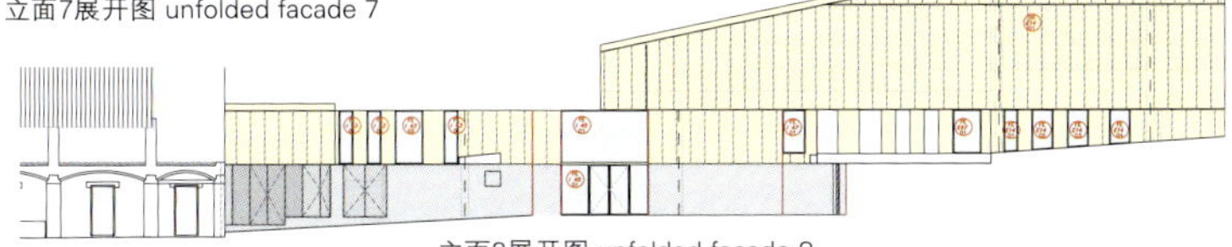
立面8展开图 unfolded facade 8

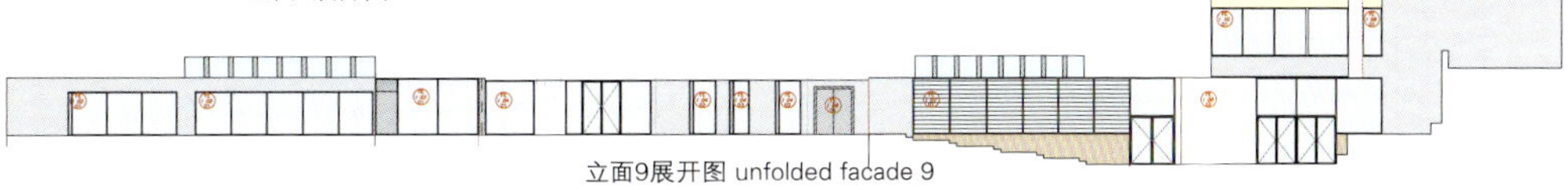
立面9展开图 unfolded facade 9

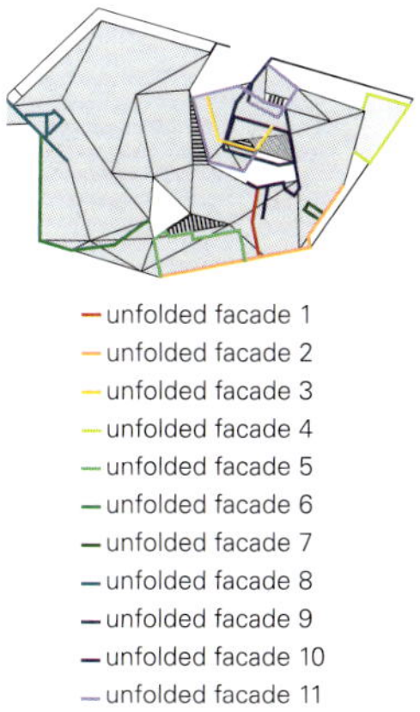

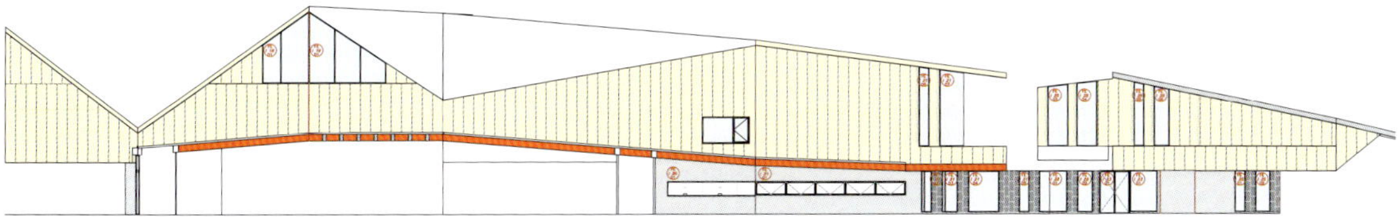
立面10展开图 unfolded facade 10

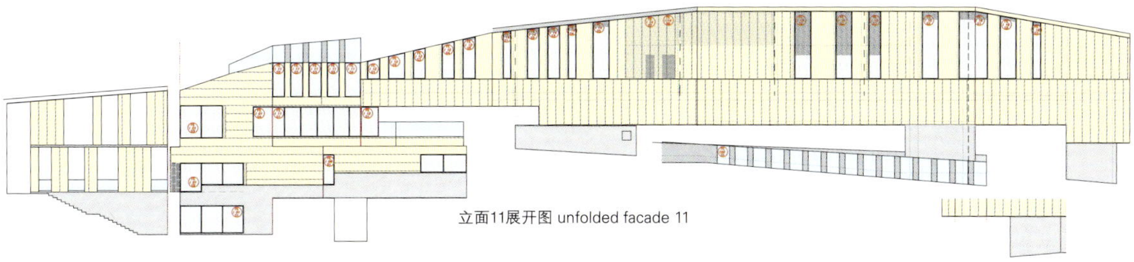
立面11展开图 unfolded facade 11

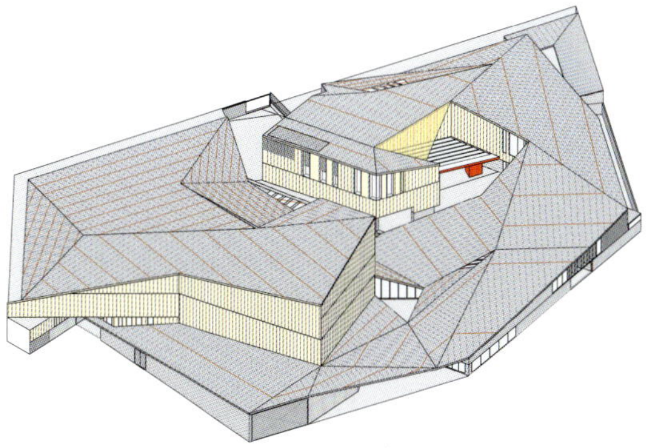

轴测图1 axonometric 1

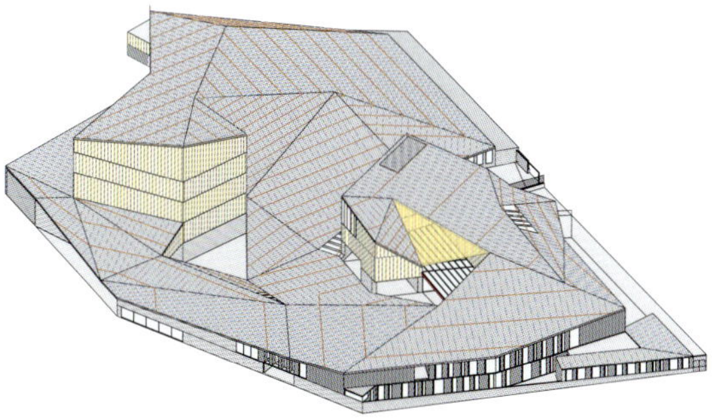

轴测图2 axonometric 2

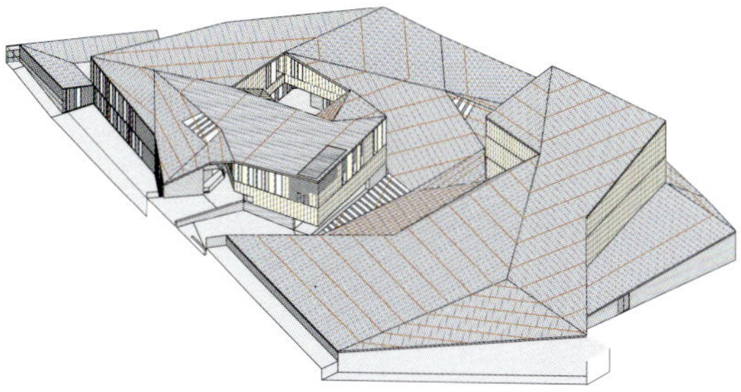

轴测图3 axonometric 3

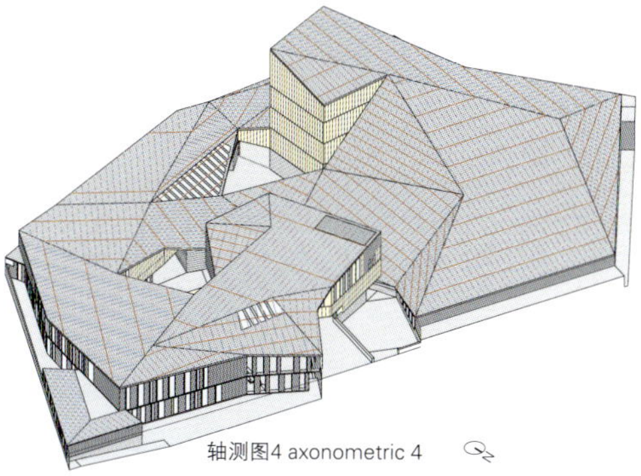

轴测图4 axonometric 4

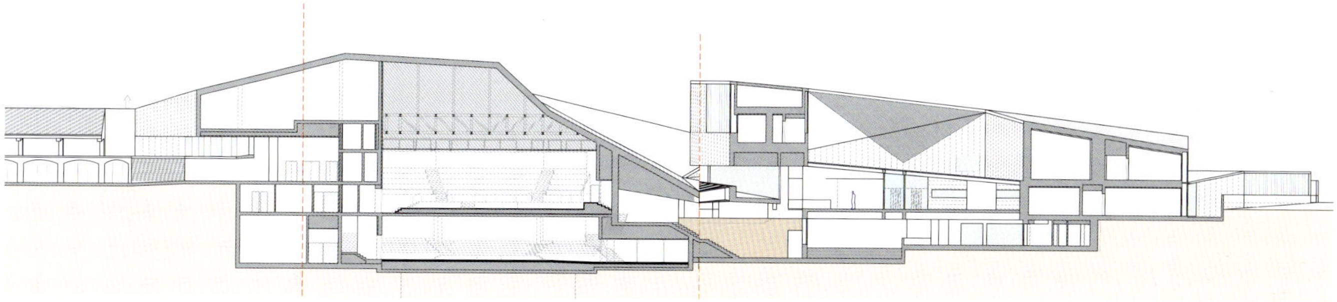

A–A' 剖面图 section A-A'

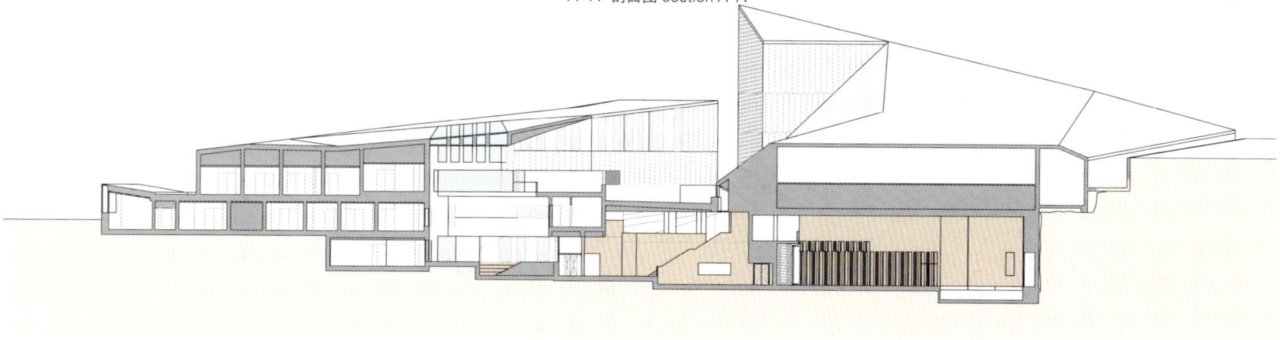

B–B' 剖面图 section B-B'

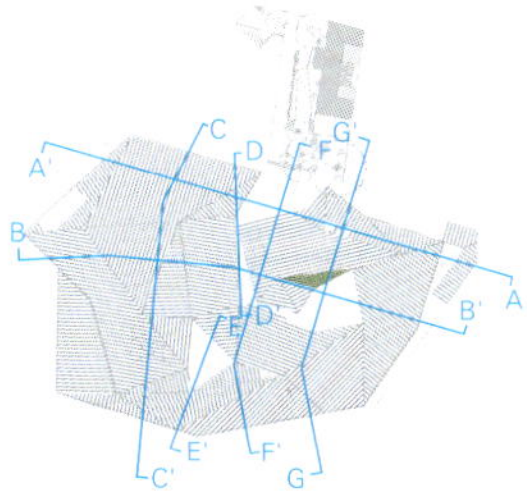

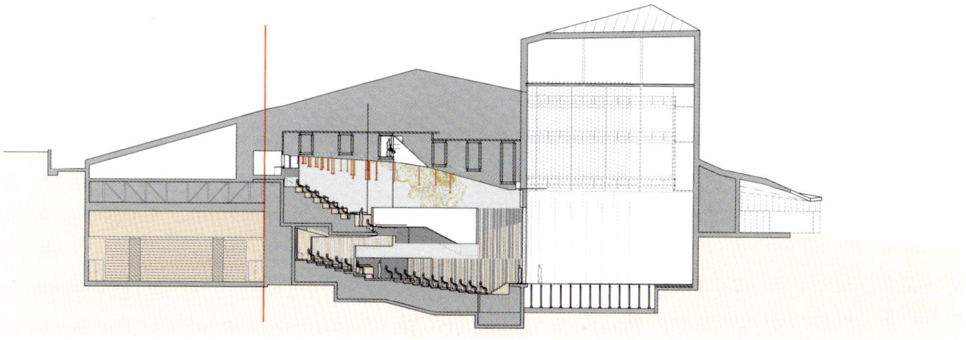

C–C' 剖面图 section C-C'

D–D' 剖面图 section D-D'

E–E' 剖面图 section E-E'

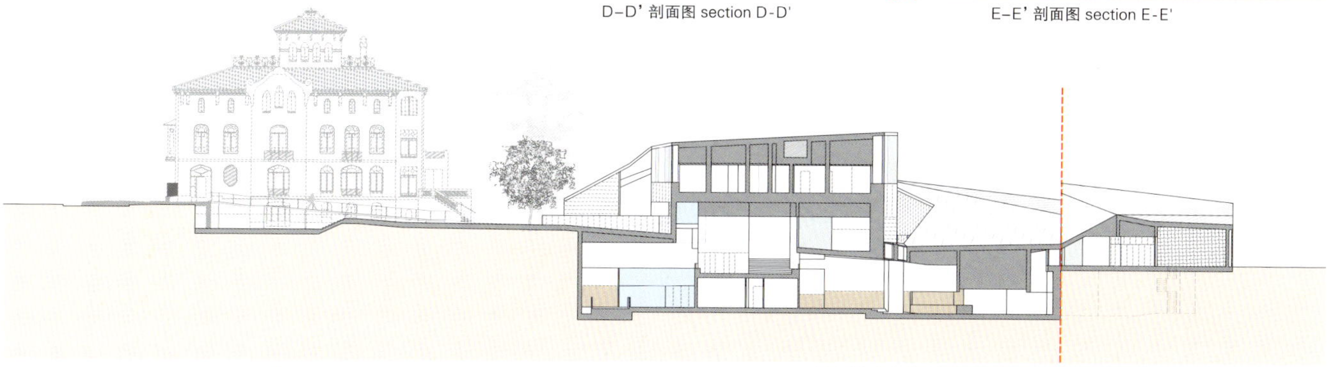

F–F' 剖面图 section F-F'

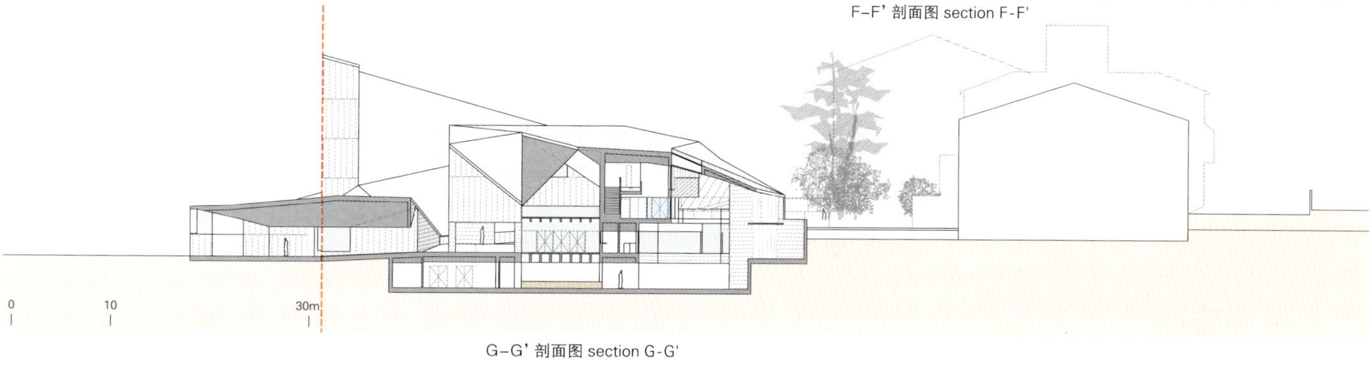

G–G' 剖面图 section G-G'

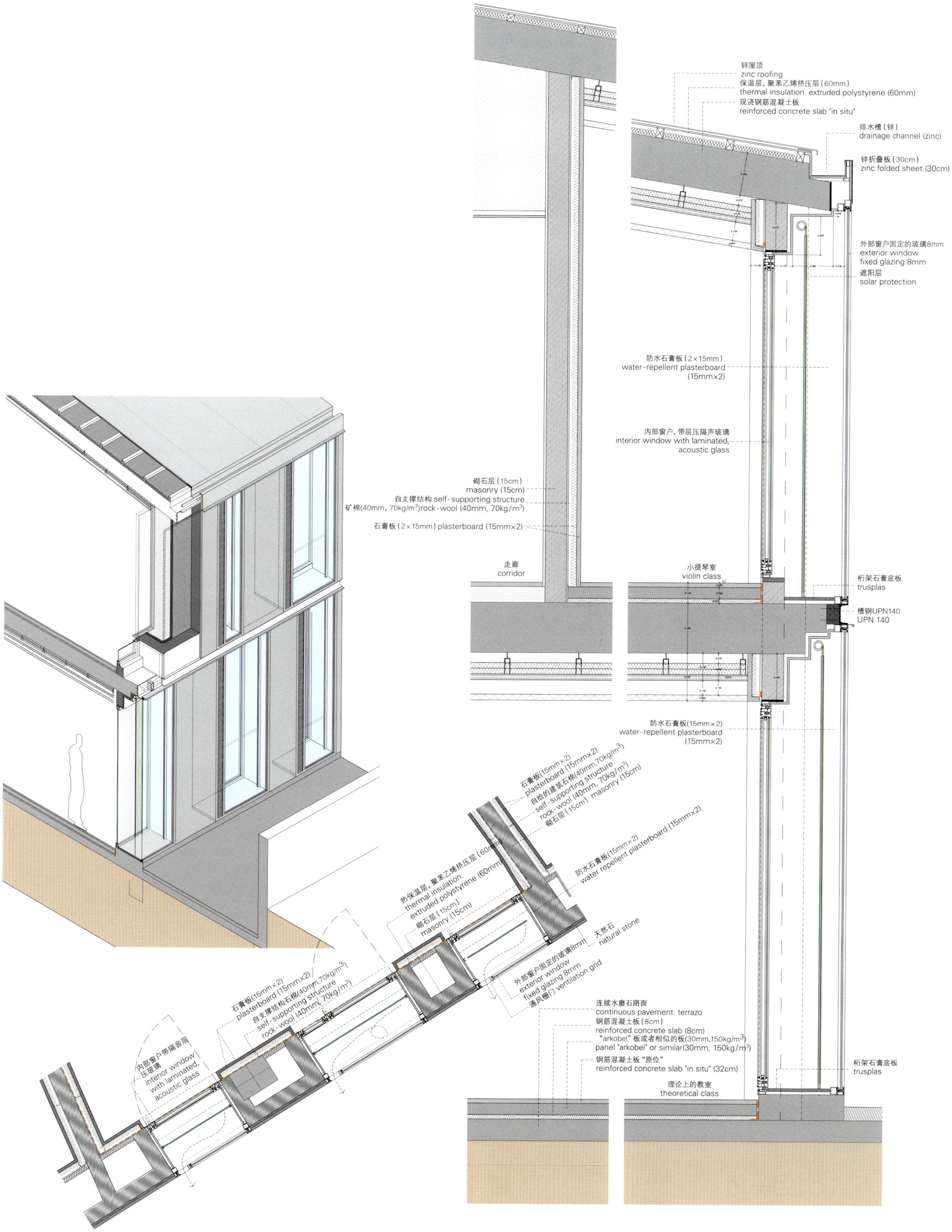
锌屋顶
zinc roofing
保温层，聚苯乙烯挤压层（60mm）
thermal insulation. extruded polystyrene (60mm)
现浇钢筋混凝土板
reinforced concrete slab "in situ"
排水槽（锌）
drainage channel (zinc)
锌折叠板（30cm）
zinc folded sheet (30cm)
外部窗户固定的玻璃8mm
exterior window
fixed glazing 8mm
遮阳层
solar protection
防水石膏板（2×15mm）
water-repellent plasterboard
(15mm×2)
内部窗户，带层压隔声玻璃
interior window with laminated,
acoustic glass
砌石层（15cm）
masonry (15cm)
自支撑结构 self-supporting structure
矿棉(40mm, 70kg/m3)rock-wool (40mm, 70kg/m3)
石膏板（2×15mm）plasterboard (15mm×2)
走廊
corridor
小提琴室
violin class
桁架石膏底板
trusplas
槽钢UPN140
UPN 140
防水石膏板(15mm×2)
water-repellent plasterboard
(15mm×2)
石膏板(15mm×2)
plasterboard (15mm×2)
自给的建筑石棉(40mm,70kg/m3)
self-supporting structure
rock-wool (40mm, 70kg/m3)
砌石层（15cm） masonry (15cm)
防水石膏板(15mm×2)
water repellent plasterboard (15mm×2)
热保温层，聚苯乙烯挤压层（60mm）
thermal insulation.
extruded polystyrene (60mm)
砌石层（15cm）
masonry (15cm)
天然石
natural stone
外部窗户固定的玻璃8mm
exterior window
fixed glazing 8mm
通风栅门 ventilation grid
石膏板(15mm×2)
plasterboard (15mm×2)
自支撑结构石棉(40mm,70kg/m3)
self-supporting structure
rock-wool (40mm, 70kg/m3)
内部窗户带隔音层
压玻璃
interior window
with laminated,
acoustic glass
连续水磨石路面
continuous pavement. terrazo
钢筋混凝土板（8cm）
reinforced concrete slab (8cm)
"arkobel" 板或者相似的板(30mm,150kg/m3)
panel "arkobel" or similar(30mm, 150kg/m3)
钢筋混凝土板"原位"
reinforced concrete slab "in situ" (32cm)
桁架石膏底板
trusplas
理论上的教室
theoretical class

项目名称："L'Atlantida", Performing Arts Center in Vic
建筑师：Josep Llinàs
相关建筑师：Josep Llobet, Pedro Ayesta, Laia Vives
技术建筑师：Miquel Autet
合作建筑师：Jorge Martín
结构计算师：Bernuz-Fernandez Arq S.L.
总工程师：GPO Ingeniería, S.A.
音响工程师：Audioscan S.L.
配景顾问：OTTO projectes S.L.
设计团队：Iván Andrés, Andrea Tissino, Philipp Gasteiger, Fermín Garrote, Iñaki Arbelaiz, Petra Pferdmenges, Natzarena Manenti, Aina Solé
甲方：Municipality of Vic. Urban Planning Department
地点：Vic, Barcelona, Spain
用途：theater, auditorium and music school
基地面积：6,500m²
建筑面积：10,500m²
设计时间：2005~2006
施工时间：2007~2010
造价：EUR 15,801,350
摄影师：©Duccio Malagamba-p.136~137, p.138, p.140, p.141, p.142, p.144, p.146
Courtesy of the architect (©FCC Construcciones)-p.135, p.136[bottom-left]

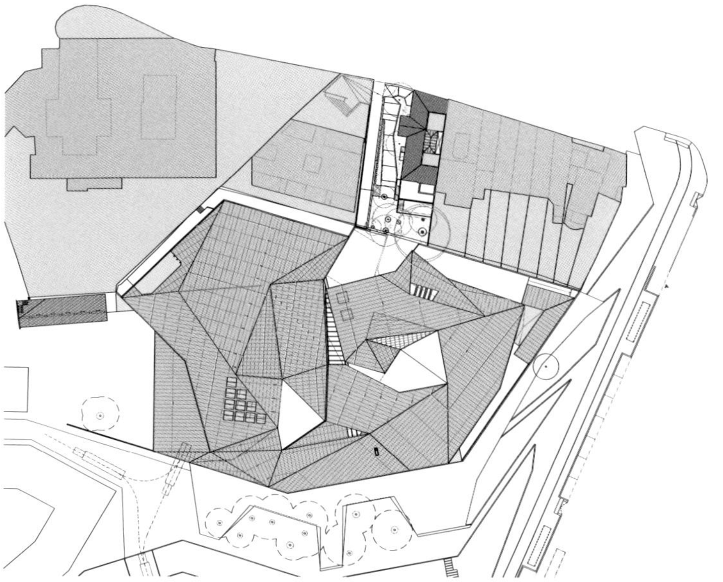
四层 fourth floor

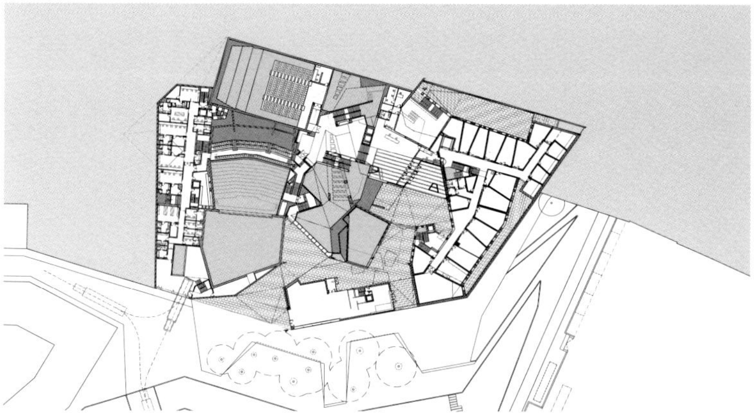
一层 first floor

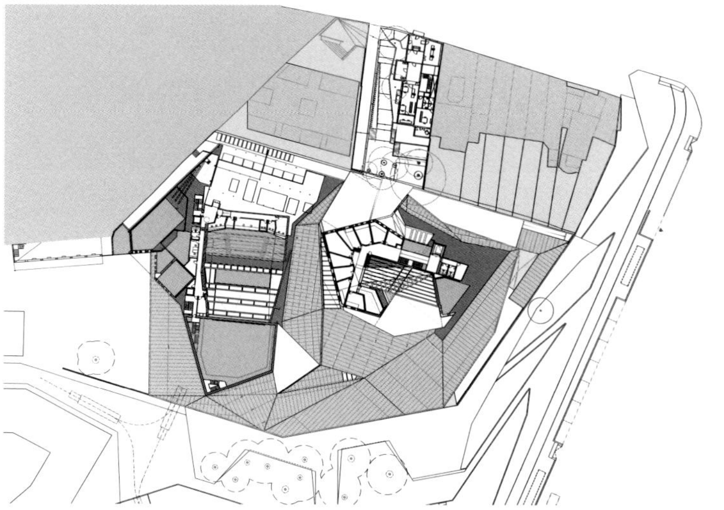
三层 third floor

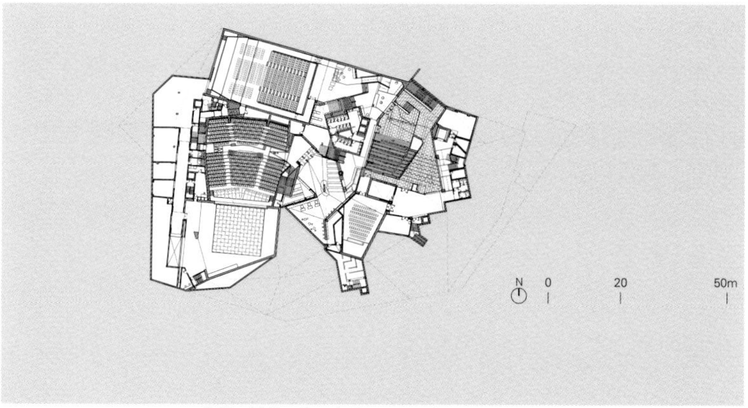

地下一层 first floor below ground

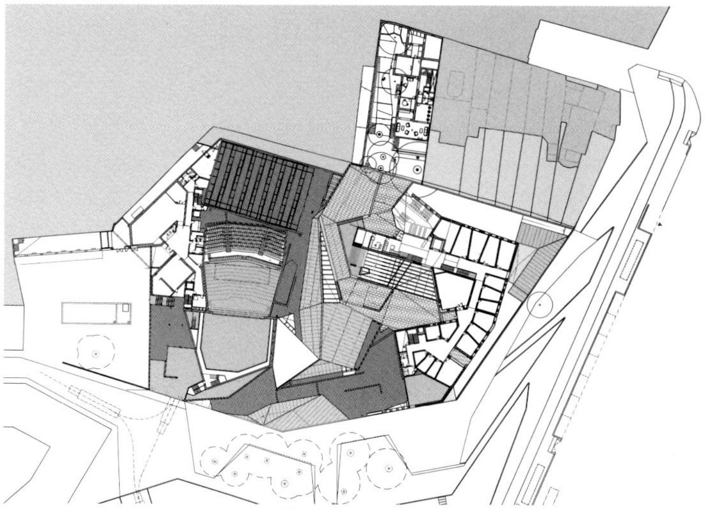
二层 second floor

扎夫拉市政剧院

Enrique Krahe

扎夫拉市政剧院坐落在西班牙西南部的一个小城镇上，其作用是协调每天不同时间举行的不同活动，并通过与行人中心的顺畅连接实现具有历史意义的城区和周边在建区域的过渡。

新建筑提供自由的略带斜坡的小路，能够让行人顺利进入，但限制车辆进入。这些由葡萄牙鹅卵石铺砌的人行道和露天花园划定了场地的范围。场地上遍布老建筑，其中两座建筑融入了本土元素（砌石拱顶和石板墙），作为将来扩大的文化活动的预留空间。

该地的不规则性需要通过消除功能性规划，创造与原有建筑规模相差无几的新的“外围护结构”，融合场地无序的几何形状，并展现一个有代表性的图像来占据场地。通过扩大由陶瓷屋顶形成的景观，主体量（包括舞台、主厅和排练厅）也得到了凸显。该设计必然会采用充分的技术手段来解决小城市对此类多功能的场所（电影院、剧院、音乐厅）的要求。但是，除了纯粹的功能性或个人喜好外，这个建筑改变了传统意义上观众的被动角色，使他们和游客、路人达到一定程度的共识，这样他们就不仅仅在表演进行的时候才能够参与进来。

几个“建筑上的手段”可以协助完成这项任务。例如，舞台（除了一个有利于直接装卸的位置之外）可以真正地向城市和市民开放。管弦乐演奏厅的座椅涂上不同的颜色，这样当座椅没有人的时候，从舞台上看，座椅会变成一只睁着的眼睛的抽象图像，这只眼睛被主空间天然毛毡装饰的墙体所环绕。

人们认为剧院应该被设计成具有激发遐想功能的场所。因此，人们已经习惯性地限制并调节了自然光的效果，并愿意使用复杂的照明控制系统，但其类型或功能的特殊性却很少对照明系统有所反应。在扎夫拉市政剧院，照明成为一个主要的议题，并或多或少地成为了工程的整体概念的基础。因此，照明具有增强项目娱乐性的积极作用，创建另一个生动的层面，引导参观者通过不太拥挤的地区，有助于创造一种安静并能够唤起情感的气氛，为人们期待的戏剧表演做筹备。

作为一个出发点，该建筑最有代表性的区域内没有使用任何商业灯具，而是创造了一组能激发情感凝聚力的照明设备，具有高性能和很强的表现力，同时缓解紧张预算并节约能源。重新利用的卫星接收器已经涂上了道路反光漆，改造成高性能、低预算的灯具，挂在门厅上方，形成了云朵一样的效果。

Municipal Theater of Zafra

Located in a small town in the southwest of Spain, the Theater of Zafra aims to reconcile the different events that are everyday staged, solving the transition between the monumental city and a “periphery” under construction, through a smooth connection with the pedestrian center.

The new organization provides free paths with gentle slopes that allow full accessibility for visitors and restricted to vehicles. Portuguese cobblestone pavements and terraced gardens delimit the perimeter of the plot, consisting of old buildings in two of which

vernacular elements (masonry vaults and slate walls) were consolidated in order to preserve them for a future enlargement of the cultural activities.
The irregularity of the plot invited to occupy the site by dissolving the functional program to create a new "envelope" with a scale close to that of the existing buildings, absorbing the disordered

项目名称：Municipal Theater of Zafra
建筑师：Enrique Krahe
项目建筑师：Aranzazu Montero
承包商：Procondal S.A.
甲方：Regional Government of Extremadura
地点：Zafra (Badajoz), Spain
基地面积：2,673m²
施工时间：2006.1~2009.11
摄影师：courtesy of the architect(©Miguel de Guzmán)

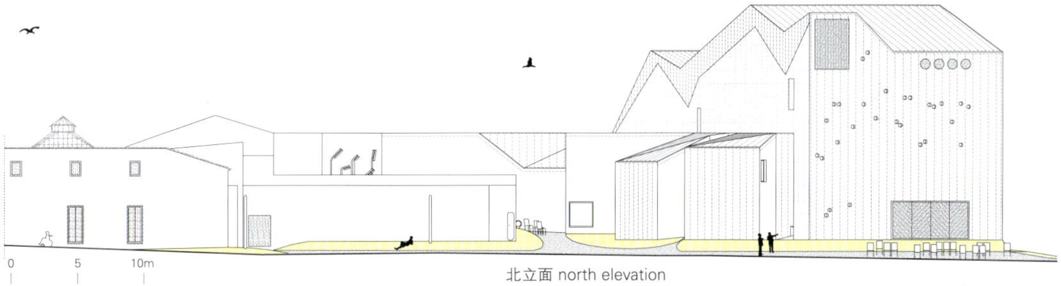

北立面 north elevation

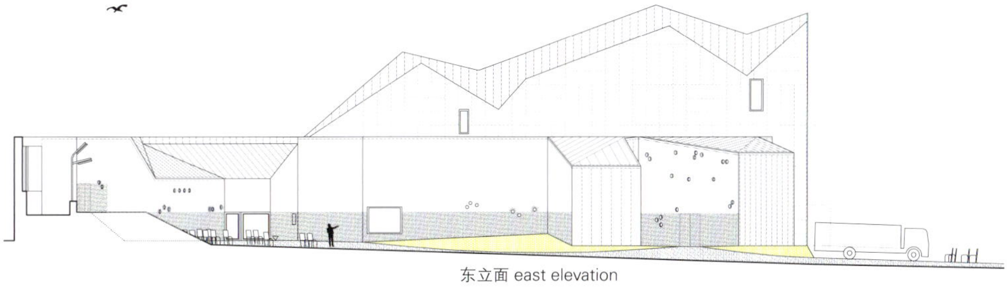

东立面 east elevation

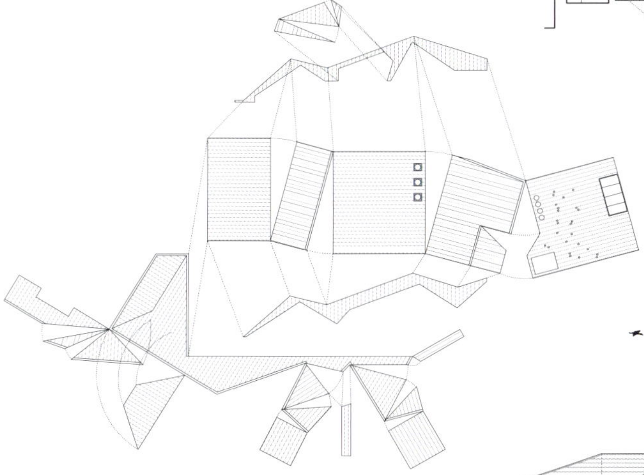

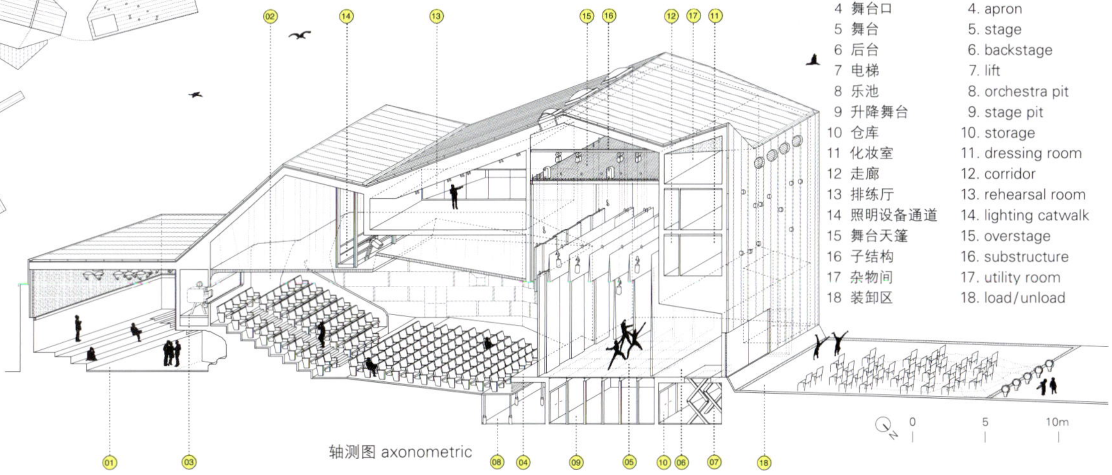

轴测图 axonometric

1 门厅	1. foyer
2 礼堂	2. auditorium
3 放映室	3. projection room
4 舞台口	4. apron
5 舞台	5. stage
6 后台	6. backstage
7 电梯	7. lift
8 乐池	8. orchestra pit
9 升降舞台	9. stage pit
10 仓库	10. storage
11 化妆室	11. dressing room
12 走廊	12. corridor
13 排练厅	13. rehearsal room
14 照明设备通道	14. lighting catwalk
15 舞台天篷	15. overstage
16 子结构	16. substructure
17 杂物间	17. utility room
18 装卸区	18. load/unload

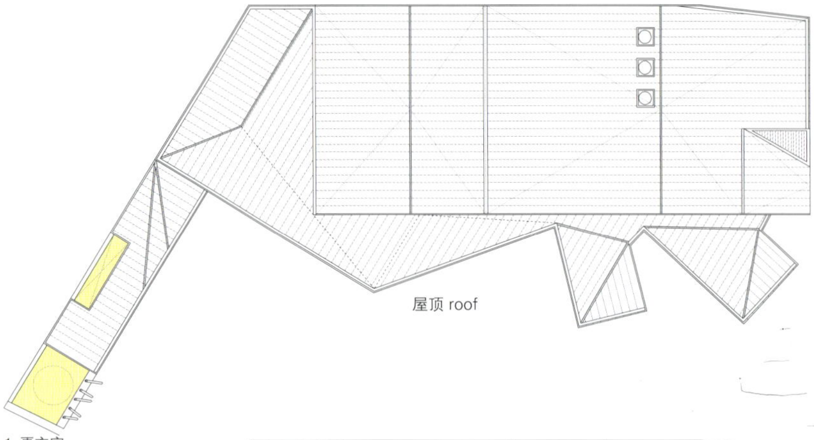

屋顶 roof

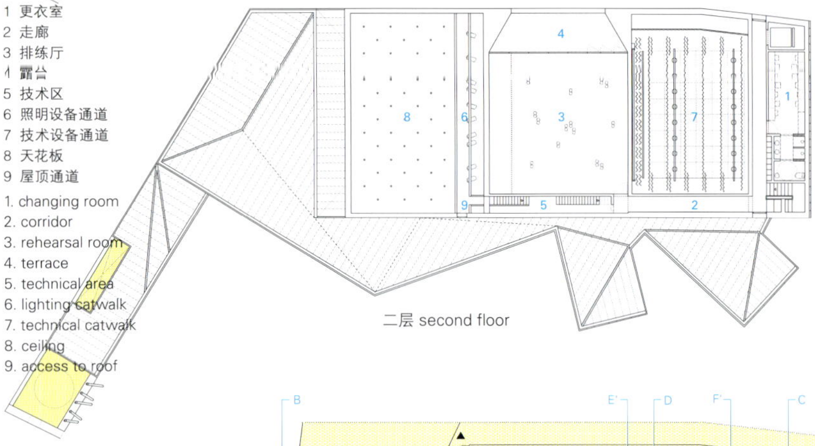

二层 second floor

1 更衣室
2 走廊
3 排练厅
4 露台
5 技术区
6 照明设备通道
7 技术设备通道
8 天花板
9 屋顶通道

1. changing room
2. corridor
3. rehearsal room
4. terrace
5. technical area
6. lighting catwalk
7. technical catwalk
8. ceiling
9. access to roof

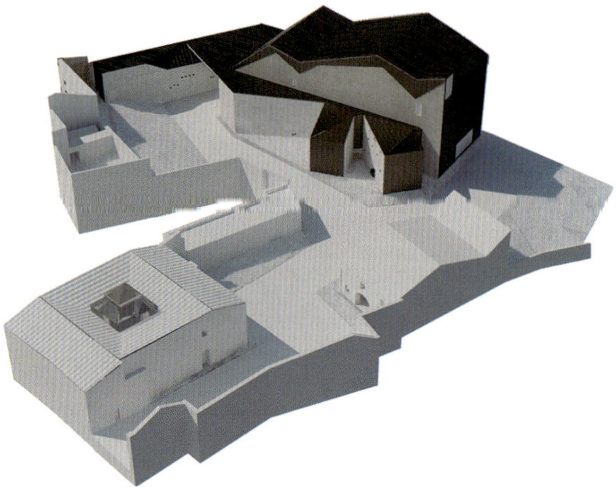

一层 first floor

1 舞台通道	1. stage access
2 主要通道	2. main access
3 演员通道	3. actors' access
4 多功能建筑	4. multi-use building
5 辅助建筑	5. auxiliar building
6 变压器	6. electrical transformer
7 现有建筑	7. existing building
8 设备	8. system
9 天井	9. patio
10 装/卸区	10. loading/unloading
11 门厅	11. foyer
12 礼堂	12. auditorium
13 售票处	13. box office
14 衣帽间	14. cloak room
15 休息室	15. rest room
16 舞台	16. stage
17 前台	17. forestage
18 后台	18. backstage
19 电梯	19. lift
20 调节器/休息室	20. dimmers/rest room
21 杂物间	21. utility room
22 通风室	22. ventilation room
23 发电机室	23. power generator

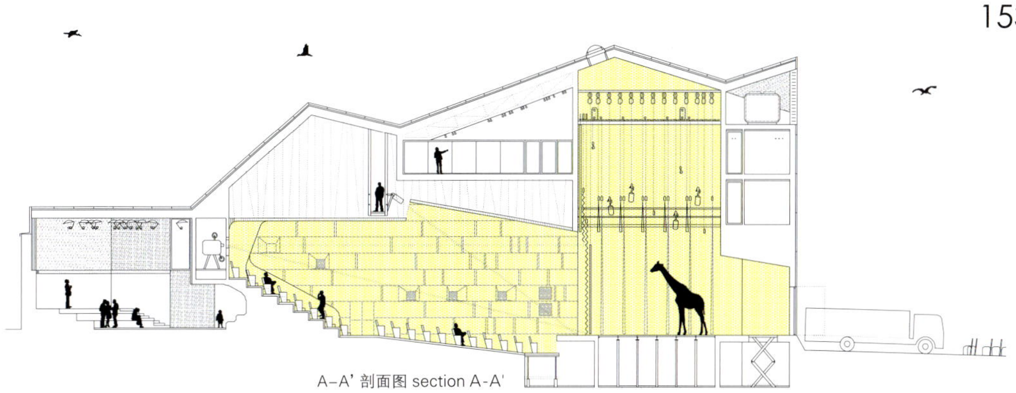

A–A' 剖面图 section A-A'

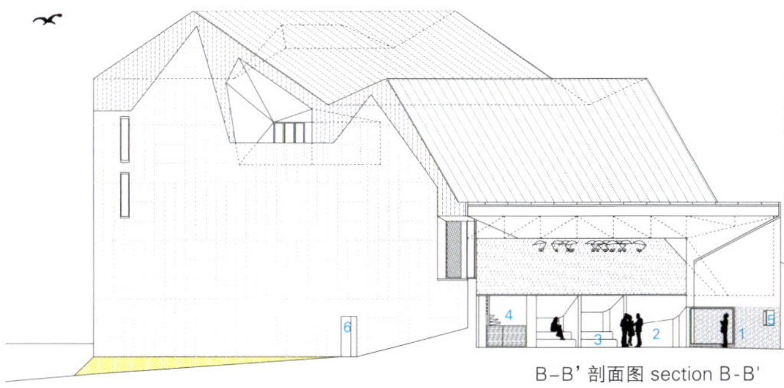

1 主入口　1. main entrance
2 门厅　2. foyer
3 专席　3. box
4 衣帽间　4. cloak room
5 售票处　5. box office
6 紧急出口　6. emergency exit

B–B' 剖面图 section B-B'

1 演员入口　1. actors' entrance
2 接待楼　2. reception building
3 舞台　3. stage
4 技术设备通道　4. technical walkway
5 通道　5. walkway
6 舞台上空　6. flyloft
7 上层舞台上空　7. upper flyloft
8 集体化妆间　8. collective dressing room
9 升降舞台　9. stage pit
10 裁缝室　10. tailoring
11 仓库　11. storage

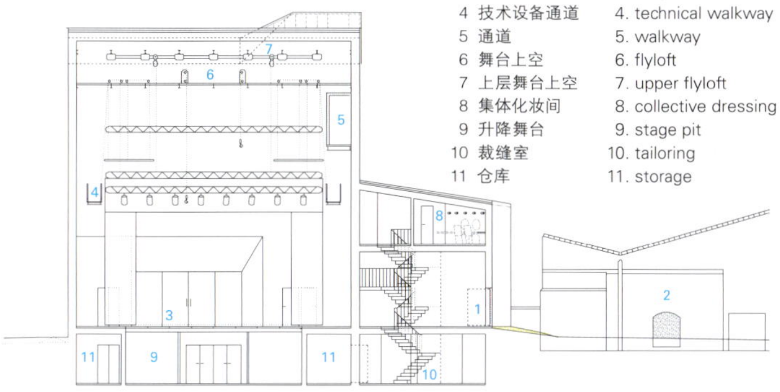

C–C' 剖面图 section C-C'

1 舞台　1. stage
2 礼堂　2. auditorium
3 排练厅　3. rehearsal room
4 屋顶　4. roof
5 技术区　5. technical area
6 技术室　6. technical room
7 机房　7. machine room
8 乐池　8. orchestra pit
9 储藏室　9. storage place
10 接待楼　10. reception building
11 露台　11. terrace
12 仓库　12. storage

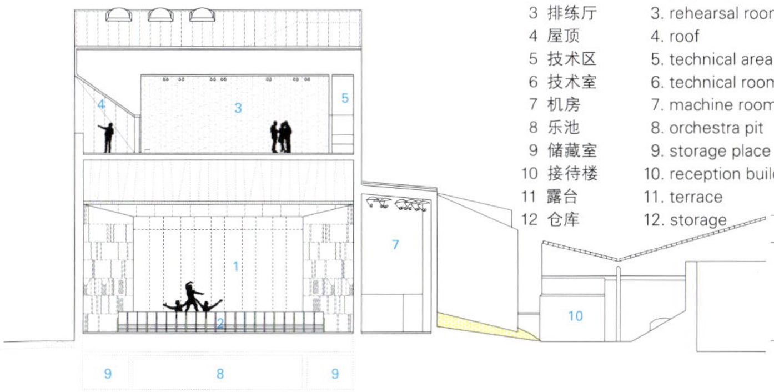

D–D' 剖面图 section D-D'

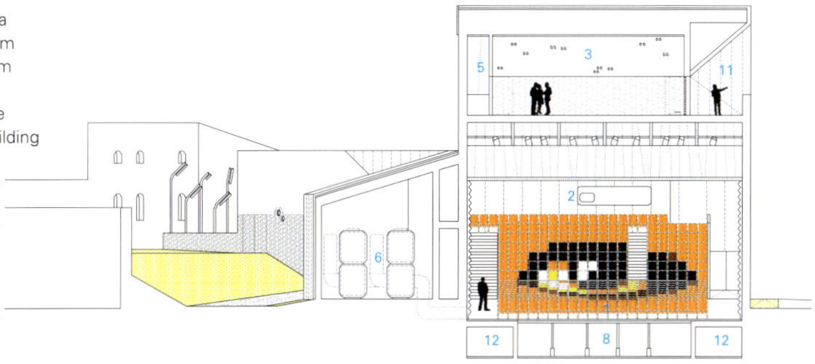

E–E' 剖面图 section E-E'

1 演员入口　1. actors' entrance
2 接待楼　2. reception building
3 舞台　3. stage
4 技术设备通道　4. technical walkway
5 通道　5. walkway
6 舞台上空　6. flyloft
7 上层舞台上空　7. upper flyloft
8 集体化妆间　8. collective dressing room
9 升降舞台　9. stage pit
10 裁缝室　10. tailoring
11 仓库　11. storage

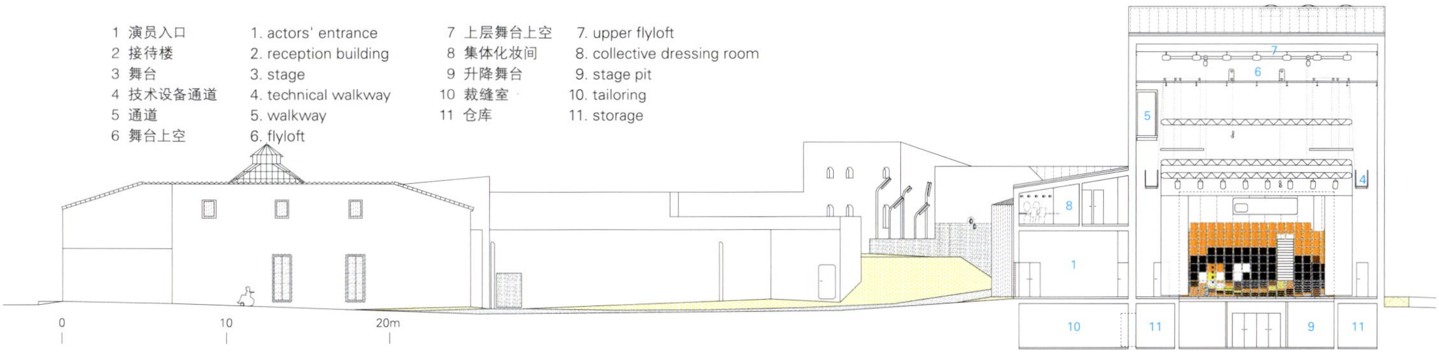

F–F' 剖面图 section F-F'

geometry and offering a representative image. The main volume (containing the box of the stage, main hall and a rehearsal room) emerges amplifying the landscape formed by the earthenware tiled roofs. The design tackles the desirable versatility of this type of facilities in a small city (film, theater, concert hall), with the inevitable technical adequacy. But beyond a mere functional or contemplative affection, the building challenges the traditional role of spectators as passive ingredients, inviting them to achieve certain complicity with both visitors and passers-by, so they can get involved not only when a representation is taking place.

Several "architectonic devices" assist on this task. For instance, the stage (in addition to offering a position that favors the direct loading and unloading) can literally open to the city and its citizens, while the seats of the orchestra are colored pixels that, viewed from the stage when empty, compose the anamorphosis of an always-looking eye surrounded by the natural felt finished walls of the main space.

Theaters are supposed to be venues functionally designed to stimulate some sort of reverie. Therefore, they have traditionally restricted and moderated the effect of natural light, in favor of a sophisticated lighting control, which rarely is fed back by its typological or programmatic particularities. At the Theater of Zafra, lighting becomes a main issue, and participates the general concepts underlying the project. Thus, it acquires an active responsibility that reinforces the playful nature of the program, creating a second graphic level, guiding visitors through the less crowded areas, and helping to create a calm and evocative atmosphere, as an accomplice preparation for the drama representation that awaits.

As a departure point, no commercial luminaires have been used in the most representative areas of the building, creating a repertoire of situations that stimulate affective bonds, all of which results in high performance and expressive effects with tight budget and energy consumption. Reused satellite dishes have been coated with road reflecting paint, and re-converted into high-performance, low-budget lamps, which form some kind of cloud above the foyers. Enrique Krahe

Local.02

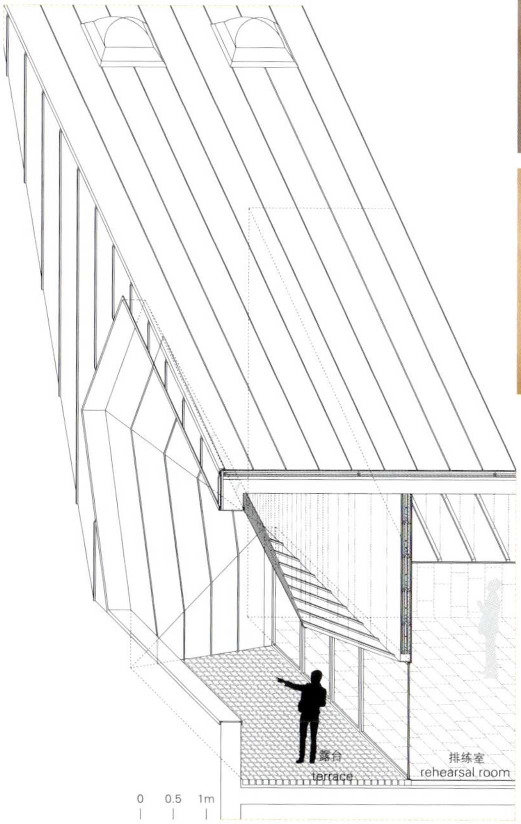

露台详图 terrace detail

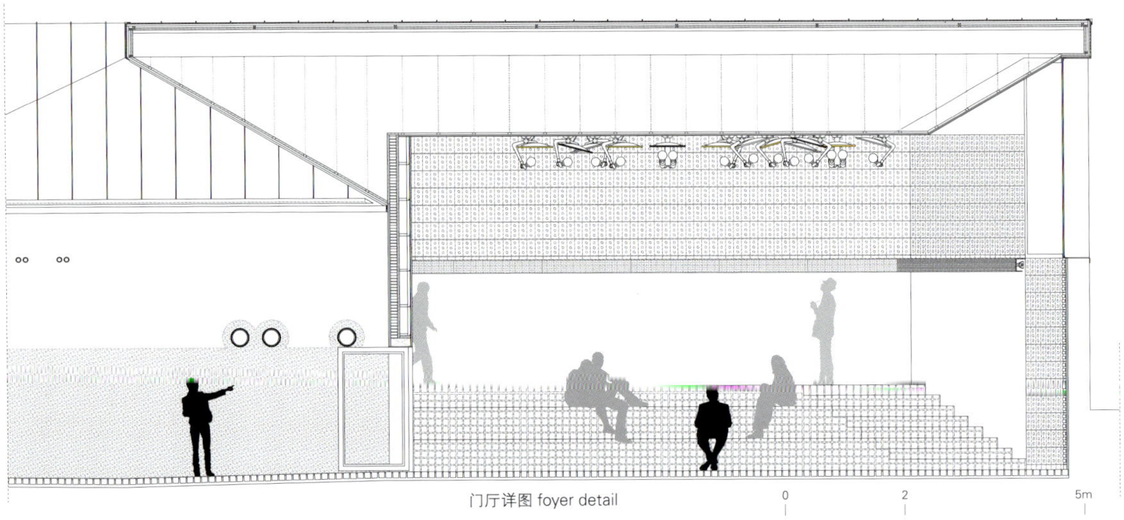

门厅详图 foyer detail

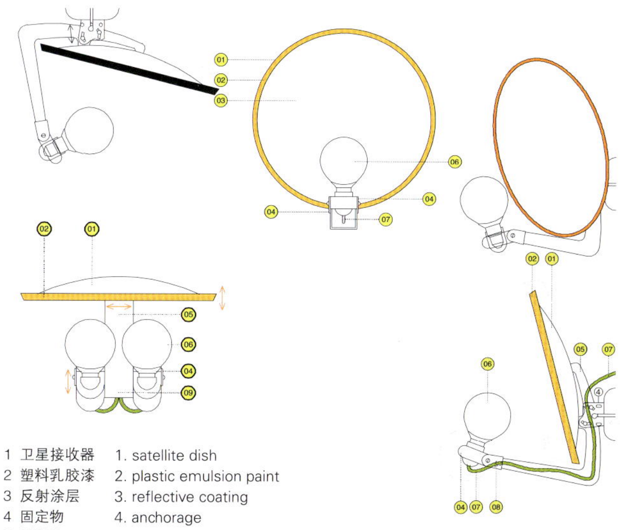

1 卫星接收器	1. satellite dish
2 塑料乳胶漆	2. plastic emulsion paint
3 反射涂层	3. reflective coating
4 固定物	4. anchorage
5 臂状物	5. arm
6 240v节能灯	6. compact lamp 240v
7 电线	7. electrical wire
8 单线接头	8. individual connector
9 双线接头	9. double connector

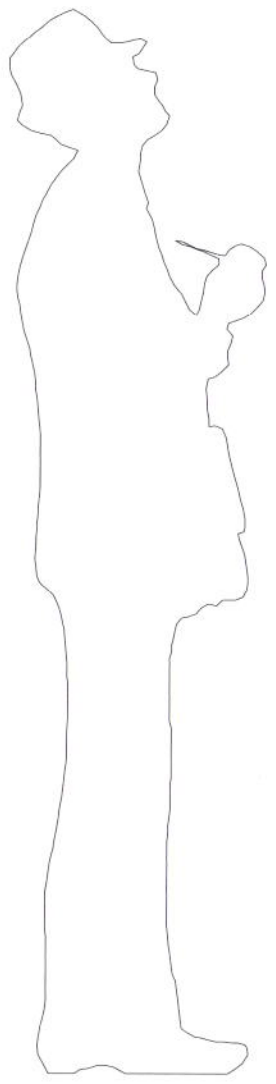

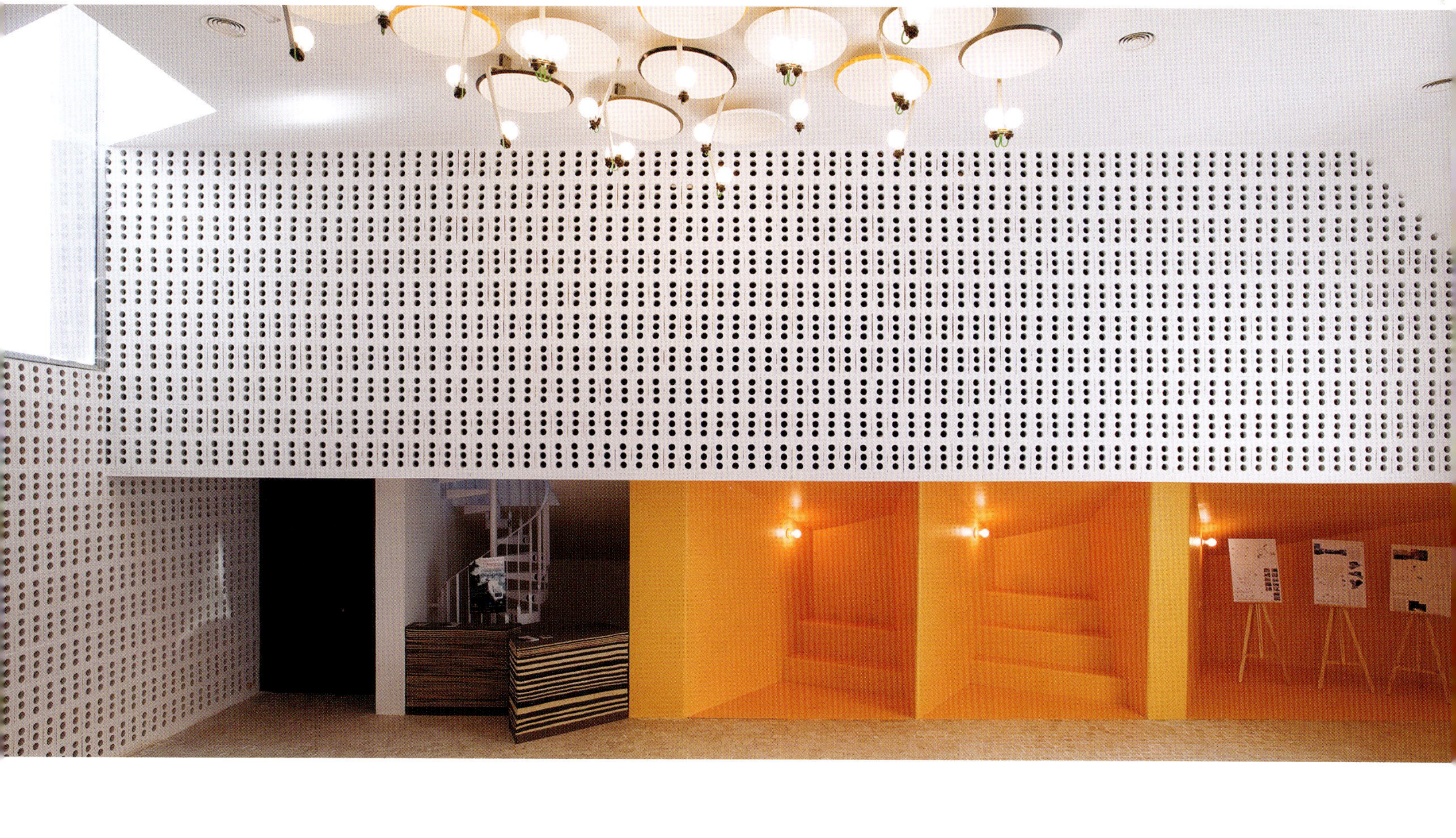

foso escenario

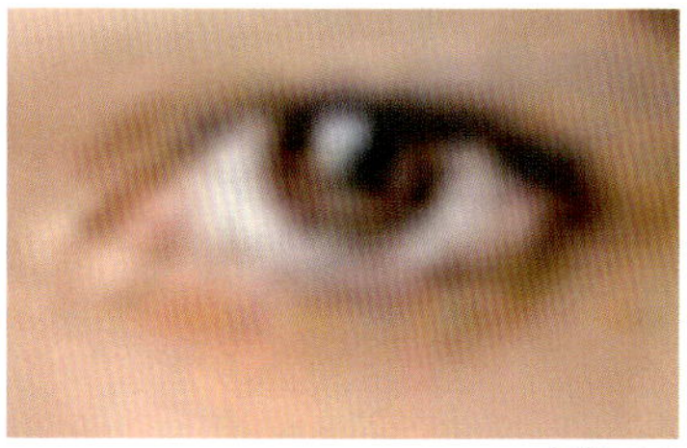

baby eye

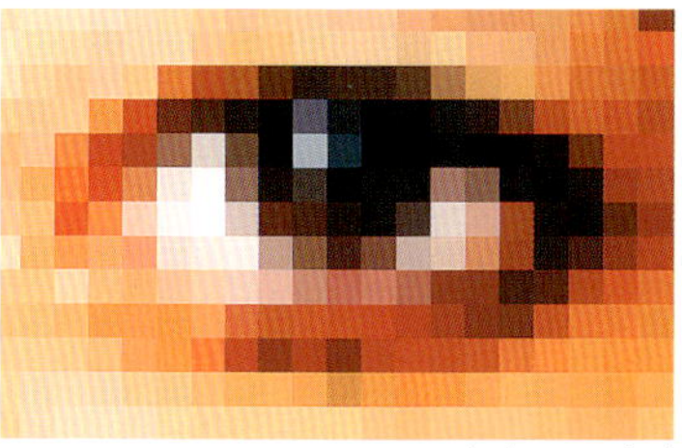

baby eye pixel

photo montage

photo book

repertoire of masks for the characterization of the public

Infanta Leonor剧院

Mrpr Arquitectos

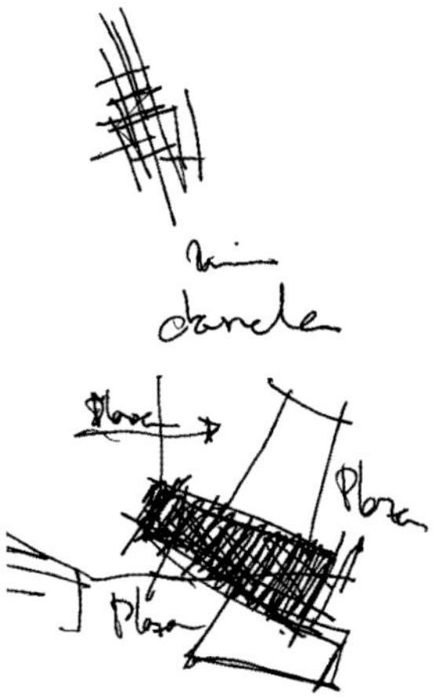

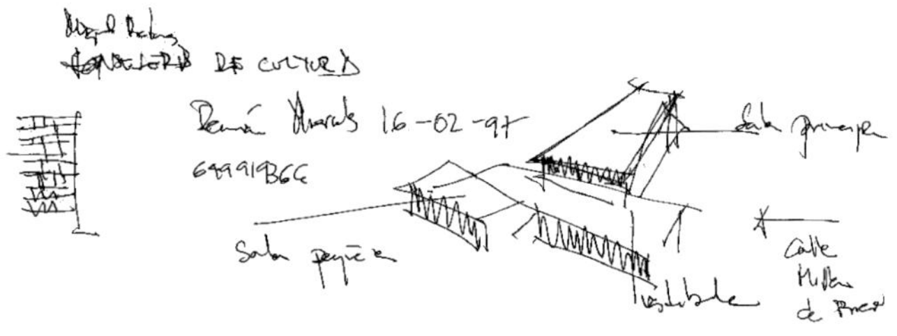

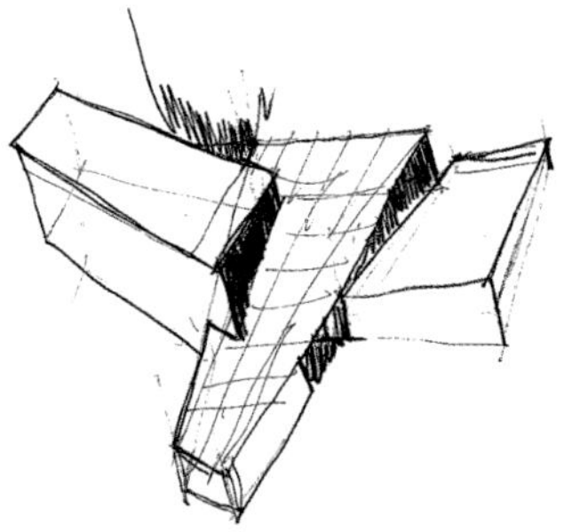

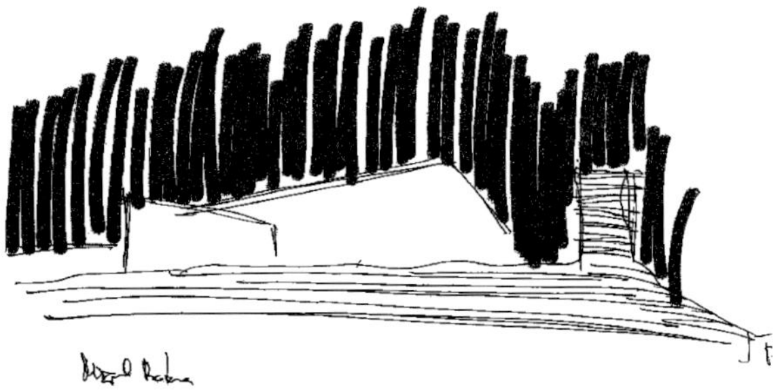

剧院主空间分为三个截然不同的部分。其占用了海拔572m高的平台，容纳了建筑的主要空间：主礼堂、大厅和排练厅。停车场、剧院办公室、化妆间、储藏室等配套设施则位于平台的下面。

从外面看，建筑的三个结构体就像三个巨大的海岬，映衬着Castillo de Santa Catalina山体，并与之交相辉映。这样设计是为了营造一系列大小不同的开放空间（广场、花园、露台）。

通过将剧院活动划分为三个区域，即主礼堂、配套空间和停车场，从而实现了建筑议程。

礼堂占用了场地的最高层，大厅位于两个礼堂之间，从大厅可以进入这两个礼堂。主礼堂包括一个有600个座位的前排座位区和有210个座位的走廊。主礼堂有舞台、乐池、格子工作台、布景储藏区和视听间。排练厅和音乐厅的舞台相对来说要小一些，只配置了容纳220位观众的前排座位、投影间和翻译间。

位于两个礼堂之间的大厅是一个巨大而狭长的错层式封闭空间。它巨大的玻璃幕墙和走廊向室外空间（广场、露台等）延伸，成功地将这些空间吸纳进来，使它们成为大厅的一部分。这个区域还配有公共卫生间、自助餐厅和衣帽间。

配套空间安排在主体结构平台的下面。演员和舞蹈演员的更衣室在海拔568m高的地方，而行政区、售票处、记者招待室、装货间和储藏室则在海拔564m高的地方。

公共停车场与配套空间在同一层，位于化妆室和办公区的分隔间后面，可容纳171部车辆。它与大厅直接相连，并通过楼梯和电梯与室外空间连接。

现场发现与哈恩的古城墙相符的考古遗址之后，这些遗址就被融入了工程中。这样有助于在以下两方面传承城市的文化遗产和历史记忆：

——直接进入遗址。

——真实地看待遗址：遗址就像一个进入点。

Puerta del Aceituno的遗址被保存并用门廊增强视觉效果，这样它成为剧院结构本身的一部分，使得售票处成为象征性的入口和第一个接触点。

因为售票处的墙壁毗邻剧院空间，所以朝向考古遗址的售票处墙面上布满了镜子，进而最大限度地扩大和彰显建筑的这一部分。

Infanta Leonor Theater

The main theater space is articulated as three clearly distinguished volumes occupying a platform at a height of 572m and housing the principal spaces in the complex: the main auditorium, the lobby and the rehearsal hall. Complementary spaces such as the car park, the theater offices, the dressing rooms, storerooms, etc., are located beneath this platform.

From the outside the building's three structural blocks appear as three enormous promontories silhouetting against and harmonizing with the hill of the Castillo de Santa Catalina. They are arranged on the site in such a manner as to create a series of open spaces of different sizes (plaza, garden, terrace).

The architectural agenda is clearly resolved by dividing the theater's activities into three areas: the main auditoriums, the complementary spaces and the car park.

The auditoriums occupy the highest part of the site, with the lobby lying between them and providing access to both. The main auditorium comprises a stall area and a gallery with 600 and 210 seats respectively. There is a stage, an orchestra pit, a gridiron, a scenery storage area and audiovisual booths. The rehearsal and music hall has a smaller stage and is equipped only with stalls, seating 220 spectators, projection and translation booths.

The lobby, located between these two auditoriums, is a large,

elongated, split-level covered space. Its large glass walls and passages project it outwards towards the exterior spaces (plazas, terraces, etc.), effectively drawing these spaces in and making them part of the lobby itself. This area also houses public toilets, a cafeteria and a cloakroom.

Complementary spaces are accommodated on the levels beneath the main structural platform. Dressing rooms for actors and dancers are at a height of 568m, while the administrative area, the ticket offices, press rooms, loading bays and storerooms are at a height of 564m.

The public car park, with a capacity for 171 vehicles, is located on the same level as the complementary spaces, behind the bay accommodating the dressing rooms and the office area. It is directly communicated with the lobby and associated exterior spaces via stairs and elevators.

Following the discovery on-site of archaeological remains corresponding to Jaen's old city wall, these remains were integrated in the project. Their inclusion contributes to recovering the city's heritage and historic memory in two ways:

– By providing direct access to the remains.

– By treating the remains, albeit metaphorically, as what they really were: a point of entry.

The remains of the Puerta del Aceituno, protected and visually embellished by a portico, thus become part of the theater structure itself, making the ticket office a symbolic place of access and the first point of contact.

Because of its proximity to the theater space, the wall of the ticket office facing the archaeological remains is covered with mirrors as a means of extending, enlarging and dignifying this part of the building as much as possible. Mrpr Arquitectos

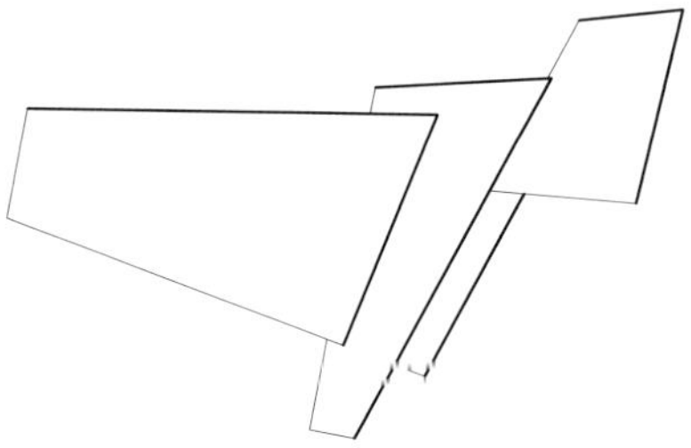

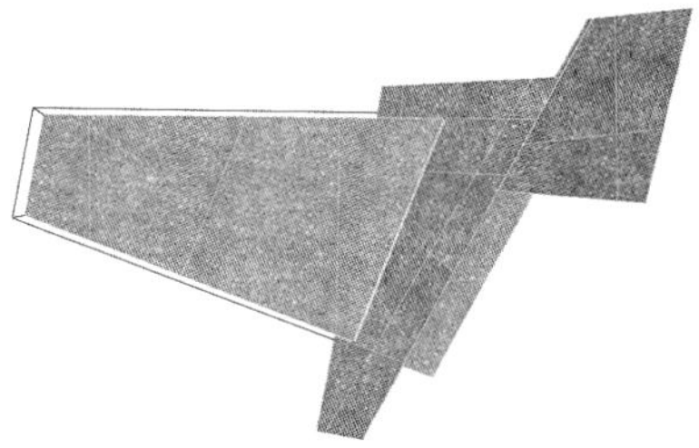

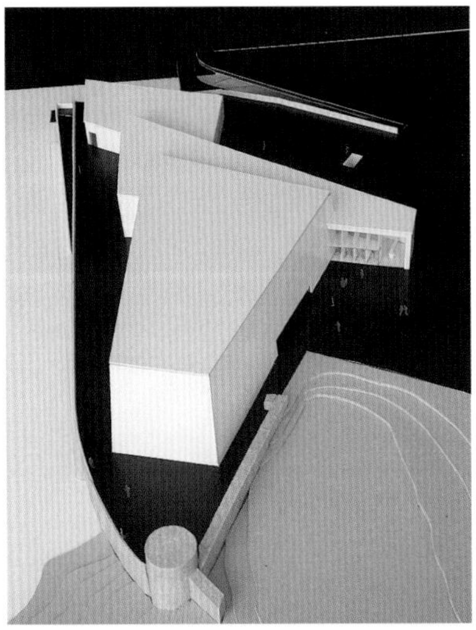

项目名称：Theater and integration of the archaeological remains of "Puerta del Aceituno"
建筑师：José Manuel Pérez Muñoz, José María Morillo Sánchez, Antonio Estrella Lara
结构工程师：Enrique Cabrera Luque. I.C.C.P.
设备工程师：Fernando García I.T.I.
空调工程师：José Luis Bocanegra Martínez de Eguilar I.T. I.
舞台设备工程师：CHEMTROL División Teatro S.A.
土木工程师：Juan Manuel González López
承建商：ACCIONA S.A.
甲方：Jaén City Council
地点：AV/Millán de Priego, Jaén, Spain
建筑面积：14,731.18m²
施工时间：2004~2008
造价：EUR 17,857,603.38
摄影师：©Jesús Granada

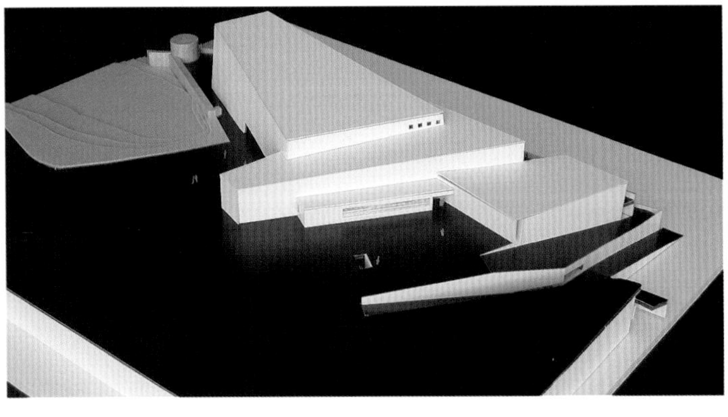

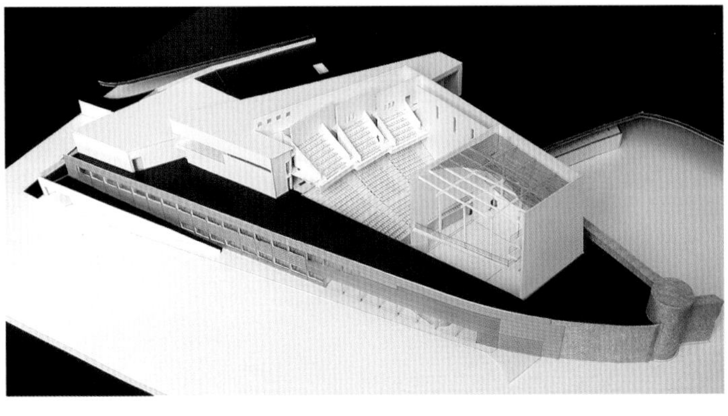

N
0
50
100m

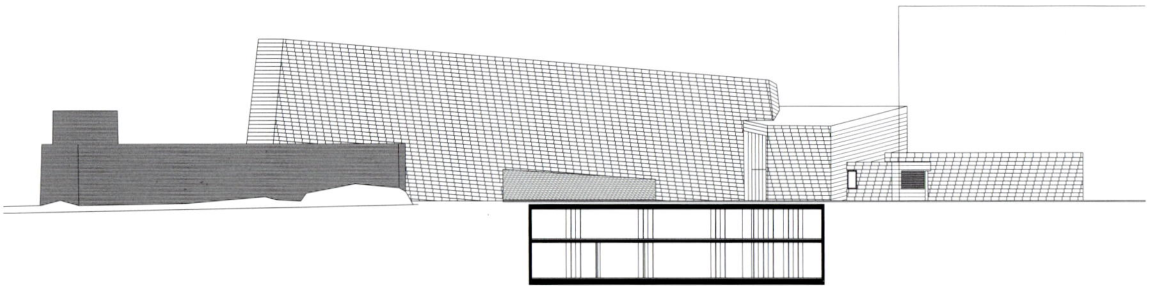

A–A' 剖面图 section A-A'

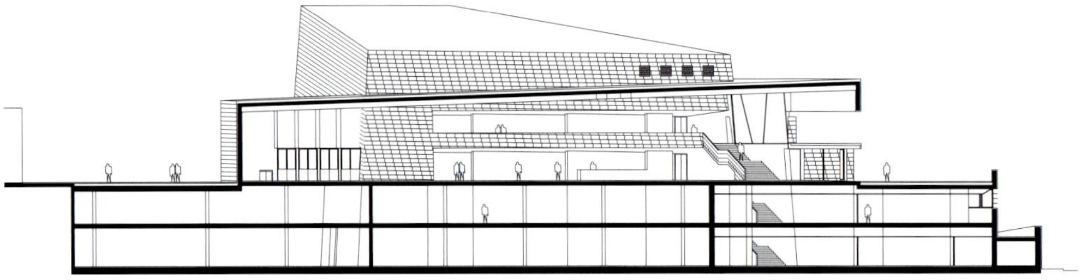

B–B' 剖面图 section B-B'

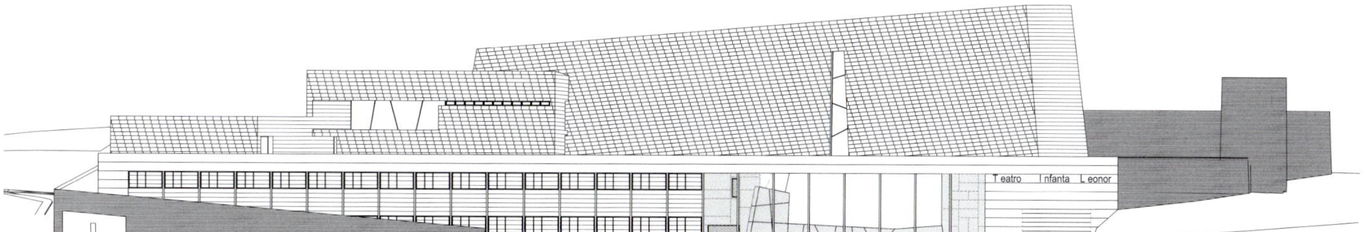

北立面 north elevation

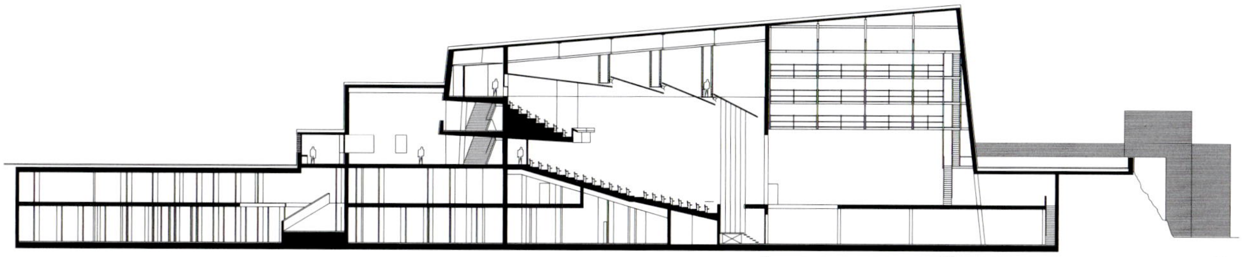

C–C' 剖面图 section C-C'

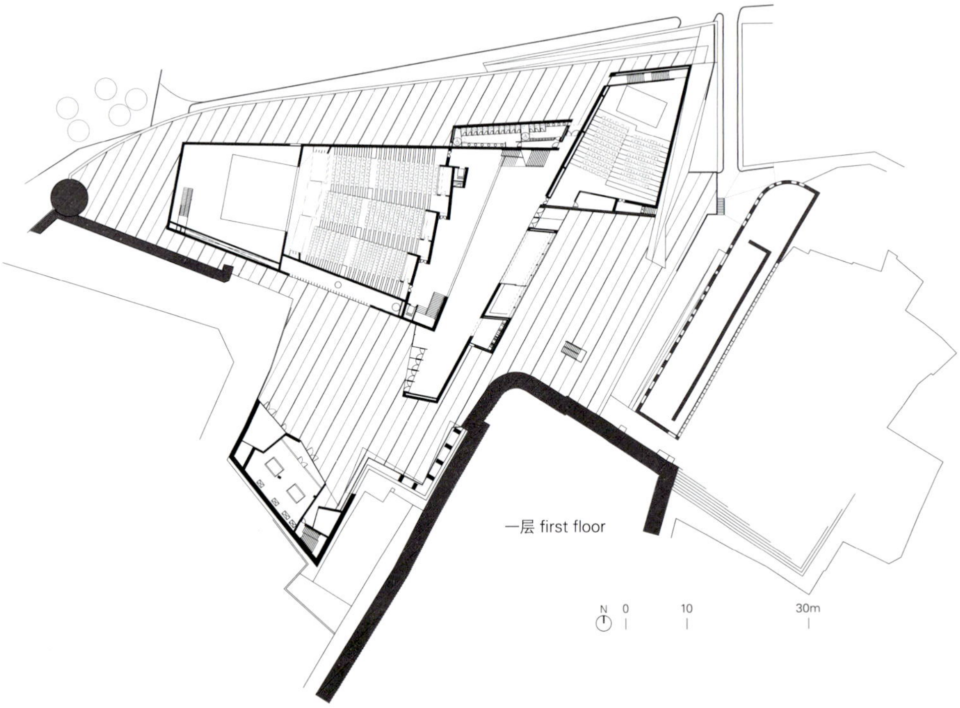

一层 first floor

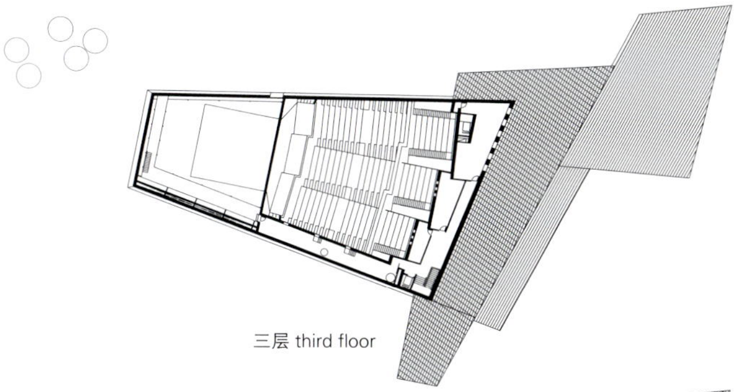
三层 third floor

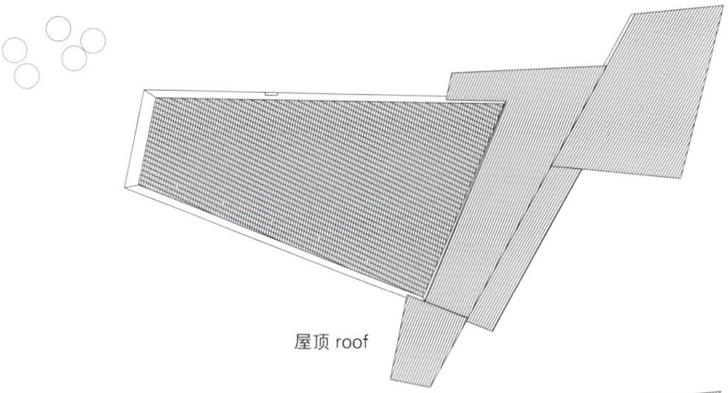
屋顶 roof

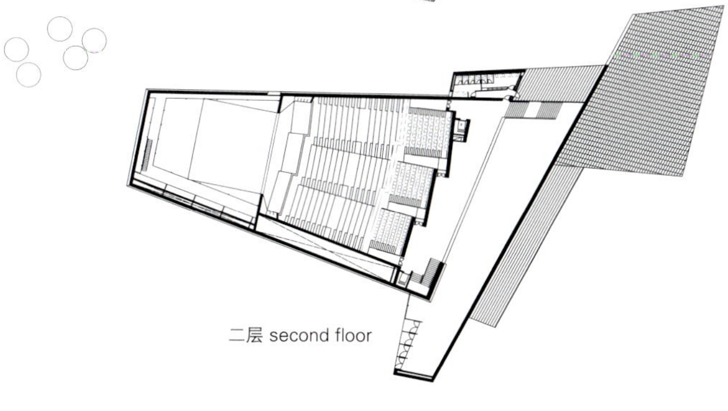
二层 second floor

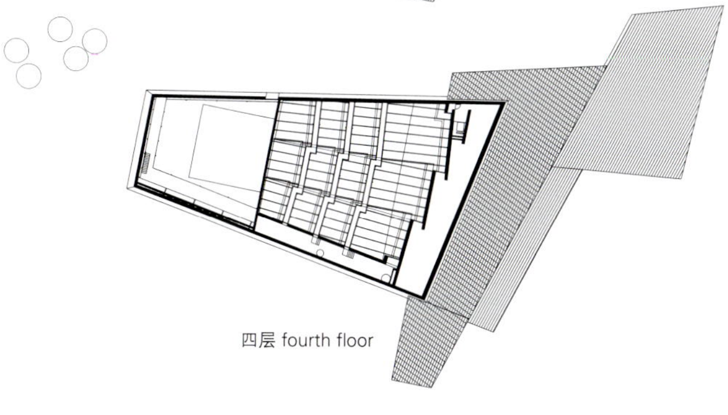
四层 fourth floor

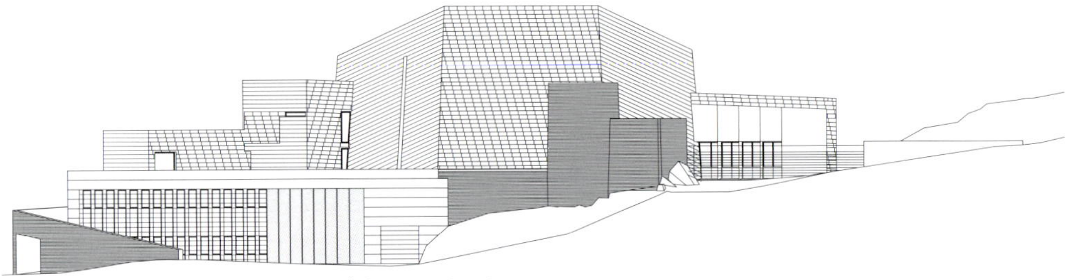

东立面 east elevation

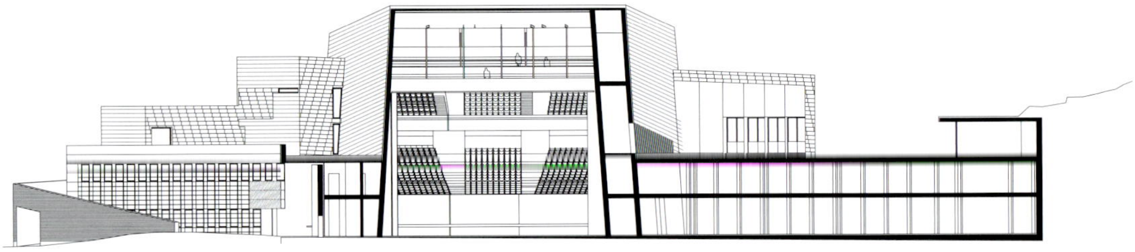

D-D' 剖面图 section D-D'

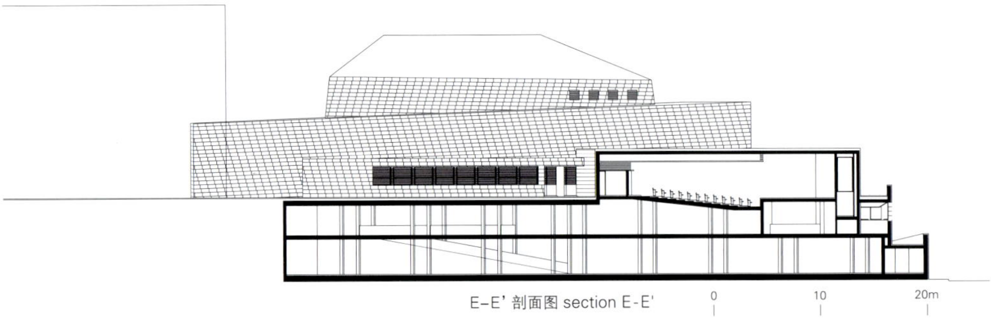

E-E' 剖面图 section E-E'

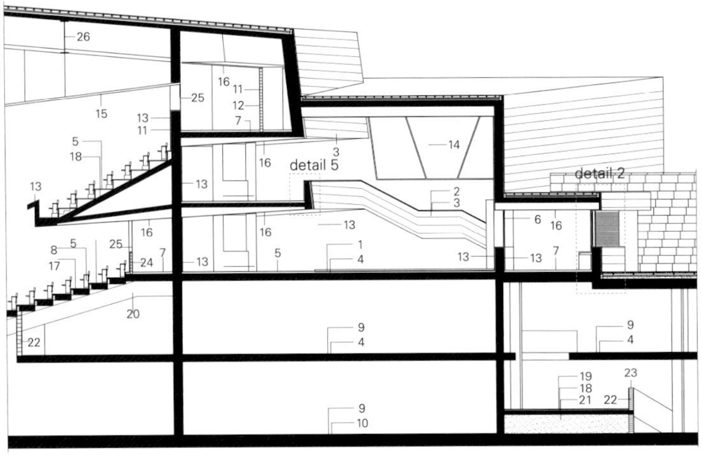

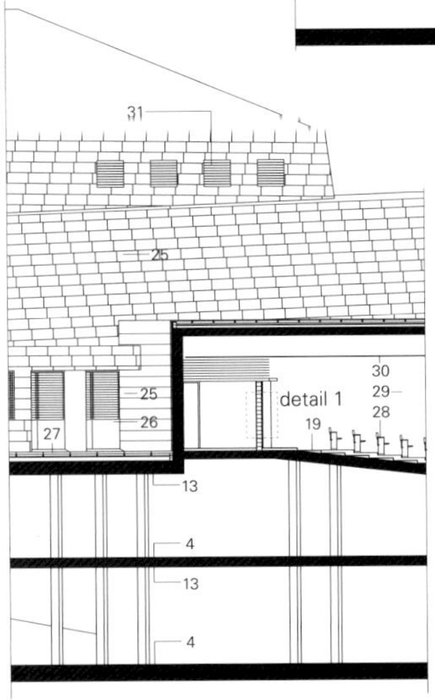

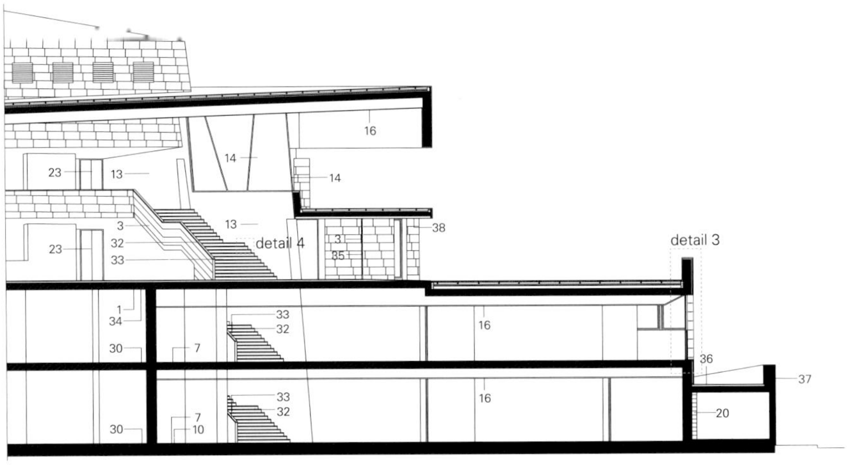

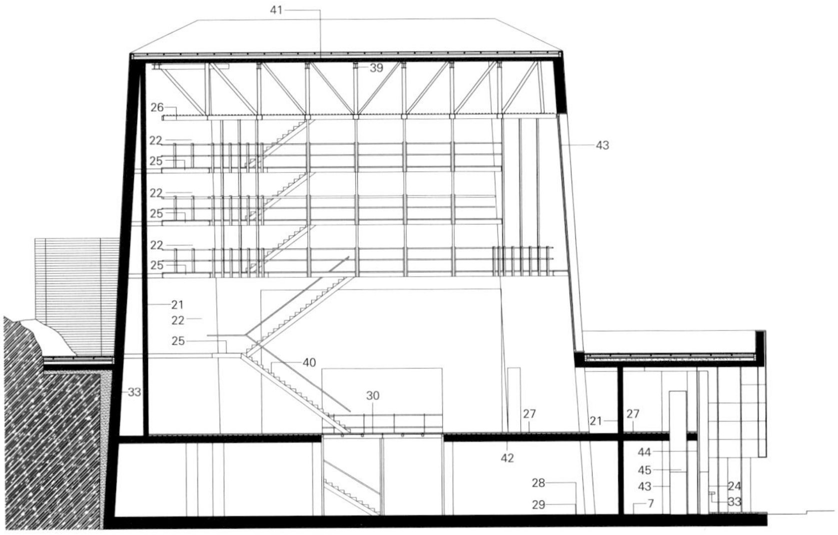

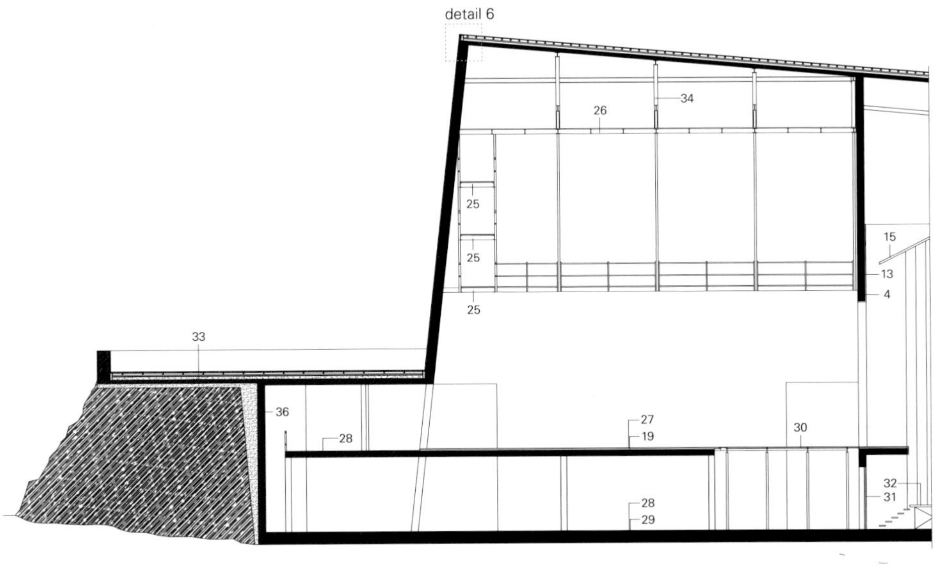

1. roman travertine flooring 60.60.3
2. U stainless steel plinth
3. clad in pieces of limestone "moka" forming a ventilated facade system, with anchors and metal substructure.
4. reinforced concrete wall ha-25
5. modular carpet
6. fixed lattice beech wood slats
7. continuous linoleum floor
8. armchair mod. maripuri
9. concrete surface treatment with aggregates
10. concrete slab
11. plaster y-12
12. extruded brick masonry 14cm
13. inner cladding formed by one fireproof gypsum panel and 12 mm thick chipboard finished with 0,5 mm beech wood layer, fixed to concrete wall or fireproof bricks through aluminium profiles
14. aluminium carpentry slide inox-nan
15. acoustic drop ceiling
16. drop ceiling
17. steps
18. ramp floor
19. reinforced concrete beam ha-25
20. filling
21. extruded brick masonry 24cm
22. cement screed m-40
23. beech wood carpentry
24. girder
25. installation service walkway
26. grid
27. melis type pine floating floor
28. surface treatment of concrete floor finish with silica, sand and gravel, corundum and quartz
29. reinforced concrete bracing slab
30. concrete floor surface finish with silica, corundum and quartz aggregates
31. metallic rf-90 carpentry
32. covering system for orchestra pit
33. gravel extrados
34. metallic truss
35. metal stairs
36. oxiasfalto primer
37. beechwood handrails scantling 300mmx50mm
38. travertine marble stairs
39. structural slab in reinforced concrete ha-25.
40. aluminium door
41. 20cm thick hollow-core deck slab
42. limestone "moka" 80.40.3
43. concrete and weathered steel plate pedestal as support for tower and wall
44. grid deck
45. pre-stress alveolar slab

1. limestone "moka" 80.40.3
2. air space
3. thermal insulation (e.g. roofmate or equivalent)
4. raised technical floor system
5. bituminous waterproofing sheet 12mm
6. 10mm cement screed
7. light concrete for gradient 600kg/m³
8. waterproofing sheet reinforcement
9. extruded brick masonry 24mmx11.5mmx7cm
10. reinforced concrete wall ha-25
11. clad in pieces of limestone "moka" forming a ventilated facade system, with anchors and metal substructure
12. polyurethane on the wall of reinforced concrete
13. drop ceiling
14. 12mm fireproof panel wrapped beech (0.5mm), fixed with anchors and metal substructure
15. metal pillars wrapped in pulished steel. 0.5mm
16. sash window. aluminium inox-nan
17. precast concrete tiles. 8cm
18. wood shelf 30mm plated in beech.
19. continuous linoleum floor
20. structural slab in reinforced concrete ha-25
21. precast concrete tiles. 8cm.
22. direct cladding in pieces of limestone "moka"
23. limestone "moka" finish
24. ceramic thin bricks board
25. extruded brick masonry 14cm
26. sandwich panel with steel cladding, insulation and aluminium frames
27. wall liner plasterboard y-25 13mm with anchors and metal substructure
28. microprofiling sandwich panel with steel cladding, insulation and aluminium frames
29. steel sheeting
30. polyurethane on the wall of reinforced concrete
31. beech handrail scantling 300mm x 50mm
32. modular carpet
33. 35w halogen lamp
34. fixed lattice beechwood slats
35. acoustic plasterboard ceiling
36. plaster y-12

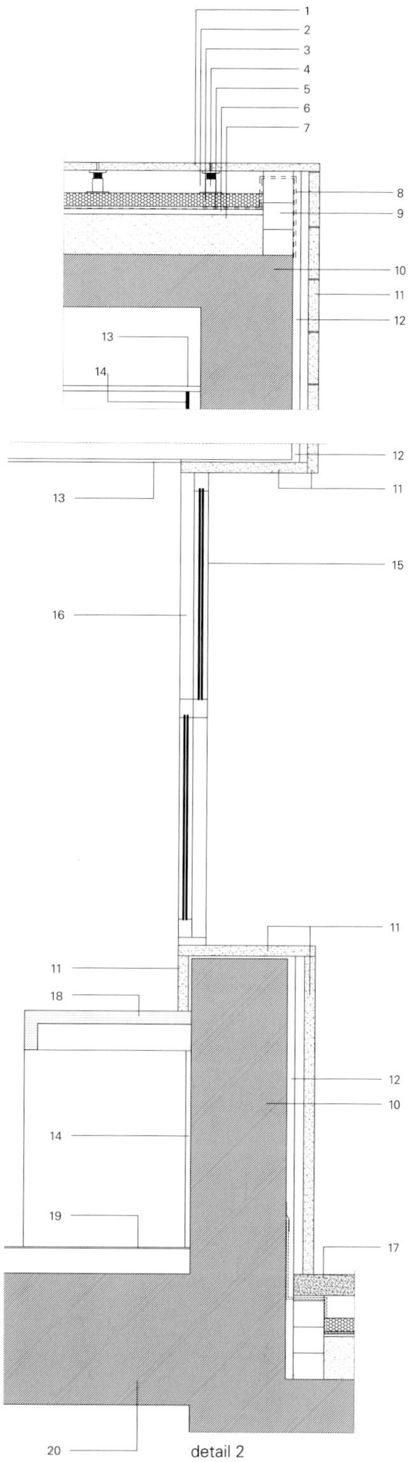

detail 2

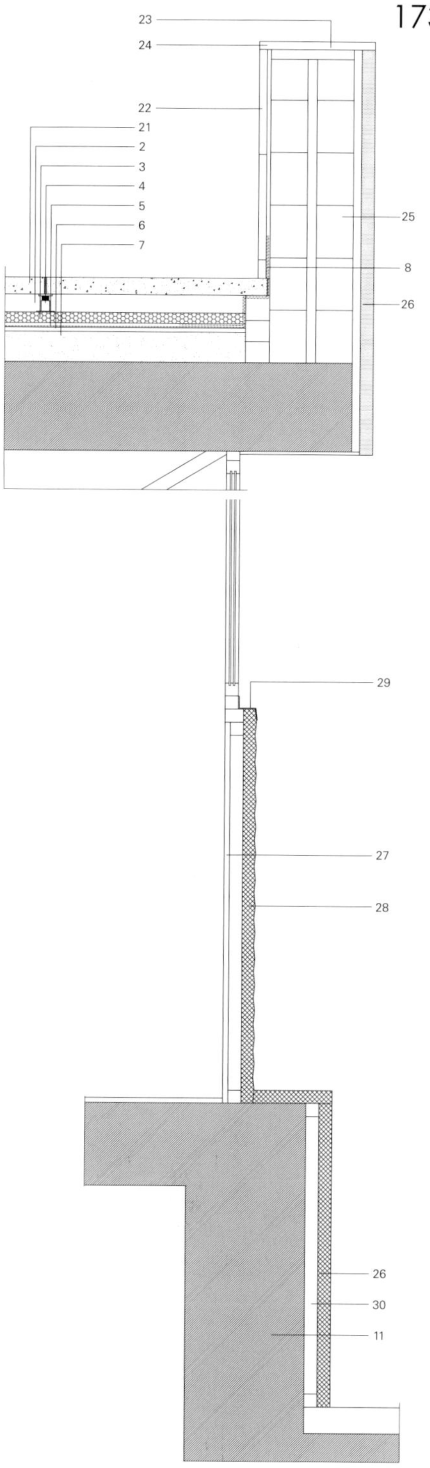

detail 3

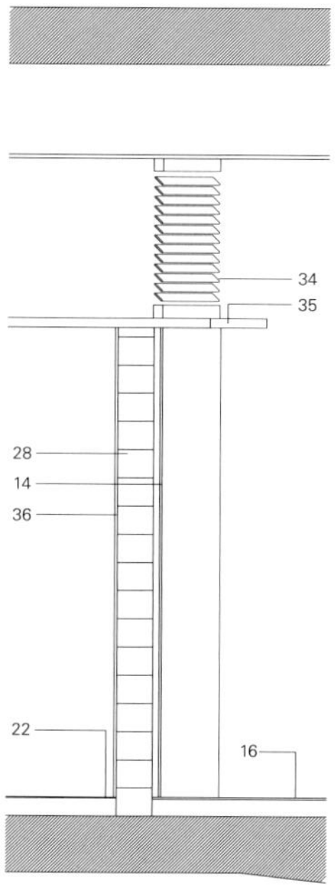

detail 1

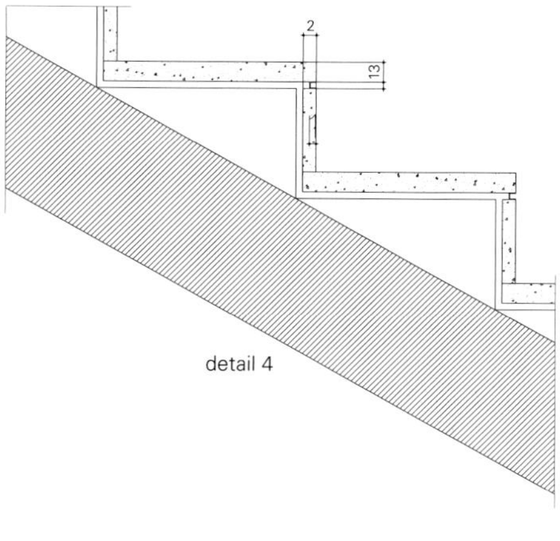

detail 4

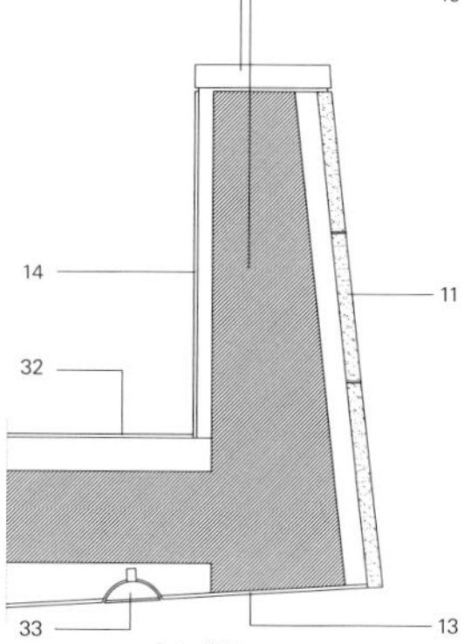

detail 5

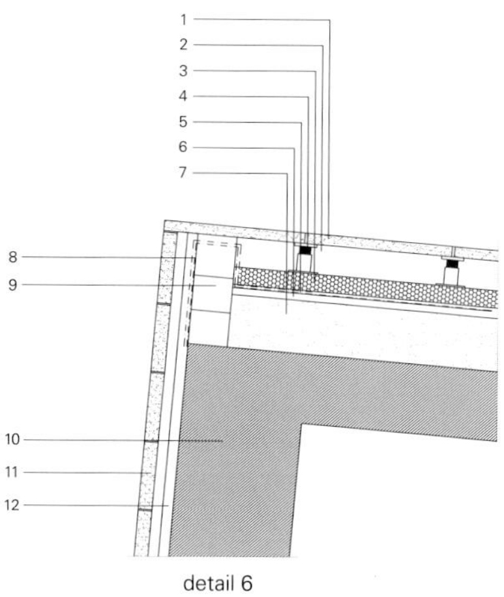

detail 6

艺术之塔

Forma 6

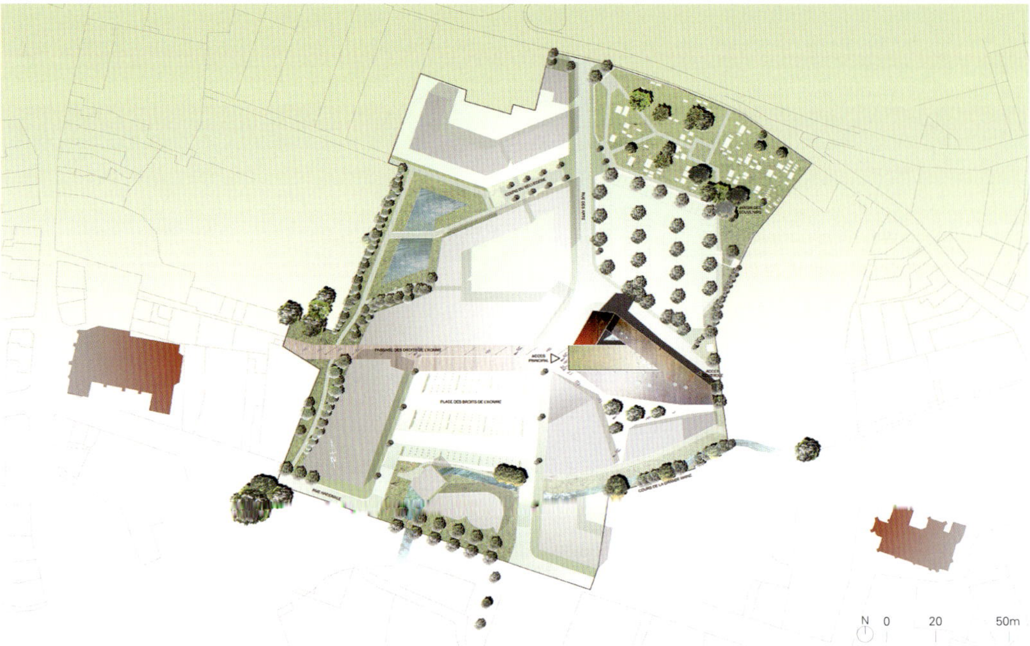

secretaria

这所音乐学校是新区内的第一幢建筑，位于市中心和郊区之间，属于人权广场的未来城市规划范畴。

它的主要房间、接待室和演讲厅与人权广场将来的人行道连成一线。它的入口与前方人行道的搭配相得益彰。其内外交界的部分自动模糊化，来增强音乐学校中心出口和人行道之间的过渡。

这座塔位于新区的前端，同时，它也是城区的标志性建筑。

学校的外观由场地复杂的几何结构和房屋的高低层次决定，学校建筑坐落于接待室和演讲厅四周，由此构成的空间既具有保护性又有活力，围绕在演讲厅周围，以这座塔作为结束。

这座塔是城市的第三个垂直标志。它与现有的钟楼交相辉映。在外形上，它具有现代气息，酷似钟楼，表层采用了与钟楼类似的金铜。选用金铜作为材料原因有二：其金色的光泽为现场的石料赋予了新的内涵；它可用于横向和纵向的表层。平面随机布局的韵律荡漾在建筑表层，就像音乐正流淌其上。

一些物质的提炼物，雕刻在这个具有保护性的金带子里，与通道和（或）自然光相呼应。这些开口通过“掉落的布块”统一进行红色标记。屋子内的自然光被垂直的窗户分散，呼应整体的布局格调。

这座工程向四周逐渐扩散，包括用金铜材料把房屋连成一个整体。这个方法更强化了建筑的特性——城镇中心和郊区的纽带。

Tower of Arts

This music school is the first building of a new district, located between the town center and the suburb, in a boundary of a future city planning of the Square of Human Rights.

The main rooms of the building, the reception and the lecture hall, are in line with the future promenade of the Human Rights.

This entrance fits perfectly with the preceding promenade. The thin limit between interior and exterior is voluntary blurred to reinforce the continuity with this promenade and the opening on the city of the music school.

The tower is in the beginning. At the same time, It is a symbolic urban signal.

The school is determined by the complex geometry of the site

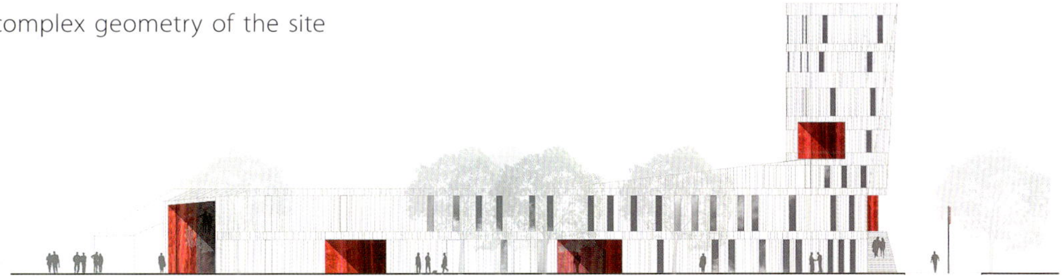

东北立面 north-east elevation

西立面 west elevation

西南立面 south-west elevation

and the various heights of the rooms and it is organized around the reception and the lecture hall. The volume created, protective and dynamic, wind around the lecture hall to finish with the tower.

The tower is the third vertical signal of the city. It acts in harmony with the existent bell towers. Its shape has been designed as a contemporary and metaphoric bell tower with its shape and its cladding, TECU Gold. This material has been chosen for several reasons: its gold color, which is a reinterpretation of the stone on the site, and its capacity to be used on horizontal and vertical cladding. The aleatory rhythm of the layout plan is vibrating the cladding as the music is vibrating the building.

Some extraction of substance, corresponding to access or/and natural lighting, is sculpting in this protective gold band. These openings are consistently marked by a red color, by a "drop cloth". The natural lighting of the rooms is diffused by vertical windows that follow the layout plan.

The ethos of the project to wind gradually around, involved treating the roof as a continuity of the material, TECU Gold. This treatment reinforced the idea of the identity of the building, a link between the town center and the suburb. Forma 6

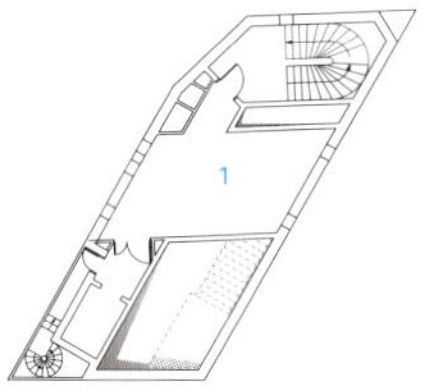

四层 fourth floor

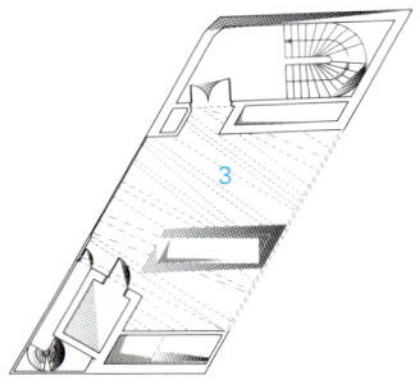

六层 sixth floor

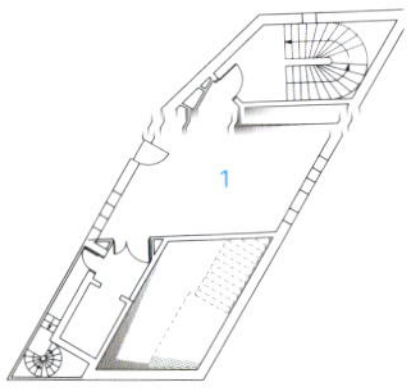

三层 third floor

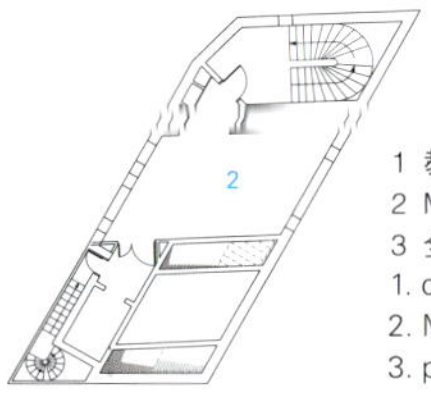

五层 fifth floor

1 教室
2 MAC电脑室
3 全景观察点
1. class room
2. MAC computer room
3. panoramic view point

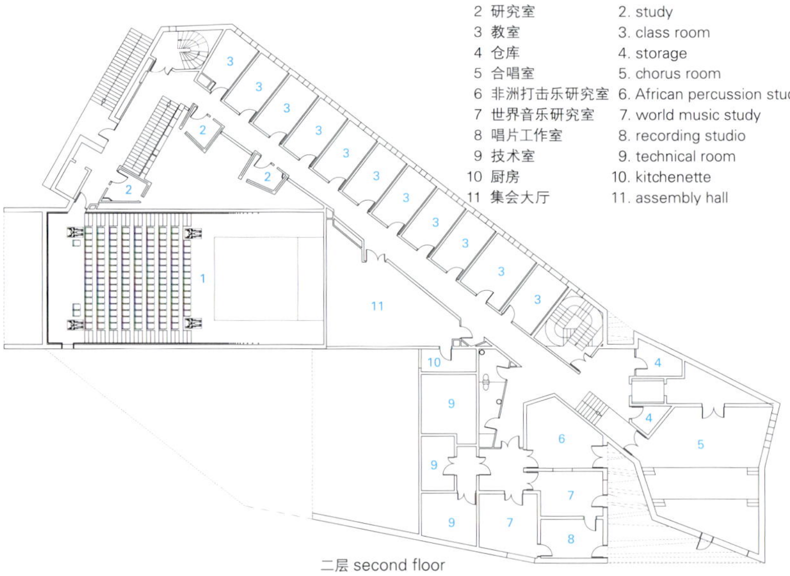

二层 second floor

1 演讲厅 1. lecture hall
2 研究室 2. study
3 教室 3. class room
4 仓库 4. storage
5 合唱室 5. chorus room
6 非洲打击乐研究室 6. African percussion study
7 世界音乐研究室 7. world music study
8 唱片工作室 8. recording studio
9 技术室 9. technical room
10 厨房 10. kitchenette
11 集会大厅 11. assembly hall

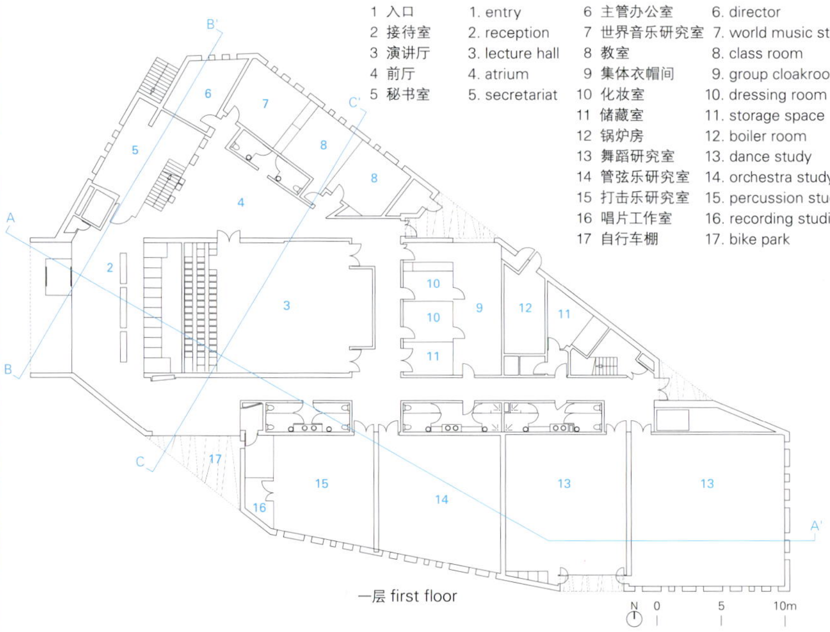

一层 first floor

1 入口 1. entry
2 接待室 2. reception
3 演讲厅 3. lecture hall
4 前厅 4. atrium
5 秘书室 5. secretariat
6 主管办公室 6. director
7 世界音乐研究室 7. world music study
8 教室 8. class room
9 集体衣帽间 9. group cloakroom
10 化妆室 10. dressing room
11 储藏室 11. storage space
12 锅炉房 12. boiler room
13 舞蹈研究室 13. dance study
14 管弦乐研究室 14. orchestra study
15 打击乐研究室 15. percussion study
16 唱片工作室 16. recording studio
17 自行车棚 17. bike park

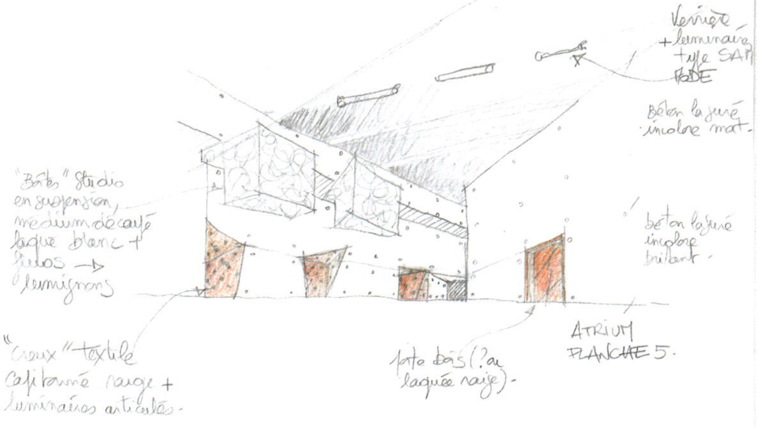
Verrière + luminaire type SAM
Béton lasuré incolore mat.
béton lasuré incolore brillant.
ATRIUM
PLANCHE 5.

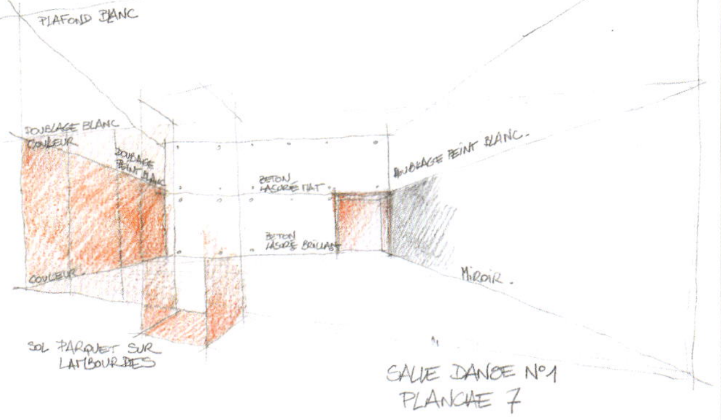
PLAFOND BLANC
DOUBLAGE BLANC
COULEUR
DOUBLAGE PEINT BLANC
BETON LASURÉ MAT
BETON LASURÉ BRILLANT
DOUBLAGE PEINT BLANC.
COULEUR
MIROIR.
SOL PARQUET SUR LAMBOURDES
SALLE DANSE N°1
PLANCHE 7

项目名称：Tour des Arts
建筑师：Forma 6
结构工程师：E2C Atlantique
机械工程师：Cabinet HAYS
电气工程师：Cabinet HAYS
甲方：Les Herbiers municipality
地点：Les Herbiers
基地面积：1,443m²
建筑面积：2,278m²
总建筑面积：1,443m²
建筑规模：five stories above ground
结构：concrete
外表面涂饰：TECU Gold
内表面涂饰：concrete
材料：concrete, fibralith, fabric, wood
设计时间：2006.2~2008.2
施工时间：2008.4~2010.4
竣工时间：2010
造价：EUR 5,234,000
摄影师：courtesy of the architect (©Patrick Miara)

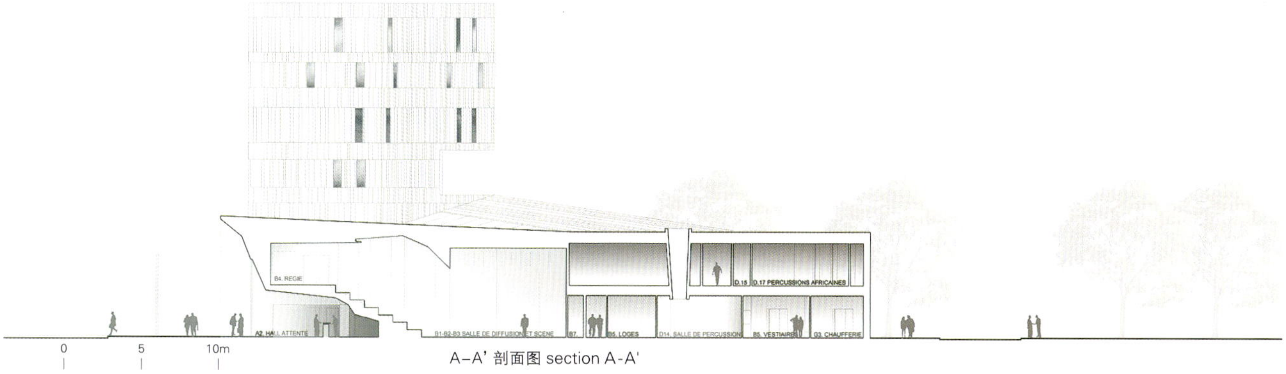

A–A' 剖面图 section A-A'

B–B' 剖面图 section B-B'

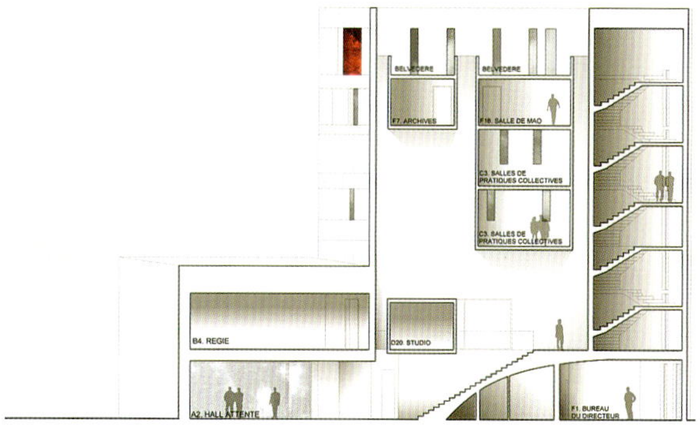

C–C' 剖面图 section C-C'

>>134

Josep Llinàs

Was born in 1945, Castellon de la Plana Spain. Graduated from Barcelona Architecture School and began lecturing at the same school. In 2005, he became a member of Alejandro de la Sota Foundation. Recently, he won the first prize of "Trophy Archizinc VM ZINC" in the category of public buildings in 2010.

>>80

Rocco Design Architects

Rocco Yim was born in Hong Kong. After graduating from the University of Hong Kong, he joined the firm of Spence Robinson for two years. In 1979, he found Rocco Design Architects. It is a Hong Kong-based architectural firm dedicated to the uncompromising fundamentals of architecture and the practice. The works are guided by a persistent attention to the quality of the total environment and sensitivity to local culture and contexts.

>>108

Alberto Campo Baeza

Was born in 1946, Valladolid. Started to study Architecture in Cadiz, Spain. He has been teaching in several Universities and Institutes. His work has been published in major architectural magazines in the world and has been exhibited in many major cities. He has built a selected number of very precise buildings. The most recent Rufo house in Toledo and Moliner house in Zaragoza, both finished in 2010.

>>92

FREE - Fernando Romero

Fernando Romero was born in 1971. He is a Mexican architect and the founder of "FREE". Its ambition is to use design to translate conditions into sustainable solutions based on economic viability as well as social and environmental integrity. In 2010, he opened his office in New York City for international operations.

>>120

Cino Zucchi Architetti

Cino Zucchi was born in Milan in 1955. Graduated from MIT in 1978 and from the Politecnico di Milano in 1979, where he is a currently chair professor of Architectural and Urban Design. Cino Zucchi Architetti is constantly searching new spatial solutions for contemporary life in the delicate and rapidly changing context of the European landscape. Its goal is to combine innovation and research with professional completeness capable of responding to complex programs on any scale and employing.

>>12

J. Mayer H. Architects

Is founded by Jürgen Mayer H. in 1996 in Berlin, Germany. Focuses on works at the intersection of architecture, communication and new technology. Jürgen Mayer H. studied at Stuttgart University, The Cooper Union and Princeton University. His work has been published and exhibited worldwide and is part of numerous collections including MoMA New York and SF MoMA.

>>162

Mrpr Arquitectos

Is composed of a group of professionals who bring together generations, attitudes and different interests in a common area of action within the framework of the Spanish production, and in particular, the Andalusian. Modern tradition inheritors, already under review, research highlights orchestrated around architecture and carried out through contrasting experiences in both existing projects and in other open formats or contests of ideas.

>>148

Enrique Krahe

Was born is 1970, Madrid, Spain. He graduated from ETSA Madrid in 1999. Since 2000 he has run his own office in Madrid, and since 2009 in Delft. Working on its own or in temporary association, his office's proposals have been awarded in several architectural competitions. At the same time, he's developed a continuous editorial activity, started in 2002 as a co-editor of Oeste Magazine. His critical essays and projects have been published in books and magazines.

>>68

Camilo Rebelo+Tiago Pimentel

Camilo Rebelo was born in 1972 Porto, Portugal. Tiago Pimentel was born in 1973, Lisbon, Portugal. Both got diploma in Architecture, Faculty of Architecture of Porto University (FAUP). Their working platform acts as an experimental field for the discussion of different classical and contemporary architectural themes. The atelier they founded wants to be an interactive platform where different realities of the contemporary condition enter in dialog in order to create a responsive architecture.

©G.Vachere

>>40

Odile Decq Benoit Cornette Architectes Urbanistes

Odile Decq set up her own office just after graduating from l'Ecole de la Villette. In 1979, she completed a post-graduate diploma in Urban Planning in Sciences Po Paris. Her first major commission was the Banque Populaire de l'Ouest in Rennes. She also designed the building for the Economic Sciences University Department and Law Library in Nantes, which was awarded a Golden Lion in Venice in 1996.

>>54

Nieto Sobejano Arquitectos

Fuensanta Nieto[right] and Enrique Sobejano[left] are graduated architects of the Escuela Técnica Superior de Arquitectura Madrid and the GSAP Columbia University, New York. They currently teach at the Universidad Europea de Madrid and at the Universität der Künste of Berlin. Both have been visiting critics at various Spanish and international universities and institutions. They won the first prize many times and got numerous awards including Aga Khan Award for Architecture with Museum & Research Center Madinat al Zahra in 2010.

>>176

Forma 6

Time, plan and the architecture projects combined with fluid mind and imagination evolve into Forma 6. Essential and collective work is the expression of approaches, methods and definitions of different contexts: social, ecological, economic, political and cultural etc. On this multiplicity of singularities and points of view, the identity of the agency is established.

图书在版编目(CIP)数据

文博建筑：中文、英文/韩国C3出版公社编；陈思，徐丽译．—大连：大连理工大学出版社，2011.10
（C3建筑立场系列丛书；6）
ISBN 978-7-5611-6568-3

Ⅰ．①文… Ⅱ．①韩… ②陈… ③徐… Ⅲ．①博物馆－文化建筑－建筑设计－汉、英 Ⅳ．①TU242.5

中国版本图书馆CIP数据核字（2011）第208817号

出版发行：大连理工大学出版社
（地址：大连市软件园路 80 号　邮编：116023）
印　　刷：利丰雅高印刷（深圳）有限公司
幅面尺寸：225mm×300mm
印　　张：11.75
出版时间：2011 年 10 月第 1 版
印刷时间：2011 年 10 月第 1 次印刷
出 版 人：金英伟
统　　筹：房　磊
责任编辑：王　培
封面设计：王志峰
责任校对：张媛媛

书　　号：ISBN 978-7-5611-6568-3
定　　价：150.00 元

发　行：0411-84708842
传　真：0411-84701466
E-mail: a_detail@dutp.cn
URL: http://www.dutp.cn